Lecture Notes in Computer Science 7050

Commenced Publication in 1973
Founding and Former Series Editors:
Gerhard Goos, Juris Hartmanis, and Jan van Leeuwen

Editorial Board

Marina L. Gavrilova   C.J. Kenneth Tan
Cong-Vinh Phan (Eds.)

Transactions on Computational Science XV

Special Issue on Advances in Autonomic Computing:
Formal Engineering Methods
for Nature-Inspired Computing Systems

 Springer

Editors-in-Chief

Marina L. Gavrilova
University of Calgary, Department of Computer Science
2500 University Drive N.W., Calgary, AB, T2N 1N4, Canada
E-mail: marina@cpsc.ucalgary.ca

C.J. Kenneth Tan
Exascala Ltd.
Unit 9, 97 Rickman Drive, Birmingham B15 2AL, UK
E-mail: cjtan@exascala.com

Guest Editor

Cong-Vinh Phan
NTT University
298A-300A Nguyen Tat Thanh Street, Ho Chi Minh City, Vietnam
E-mail: pcvinh@ntt.edu.vn

ISSN 0302-9743 (LNCS)   e-ISSN 1611-3349 (LNCS)
ISSN 1866-4733 (TCOMPSCIE)   e-ISSN 1866-4741 (TCOMPSCIE)
ISBN 978-3-642-28524-0   ISBN 978-3-642-28525-7 (eBook)
DOI 10.1007/978-3-642-28525-7
Springer Heidelberg Dordrecht London New York

Library of Congress Control Number: 2012931776

CR Subject Classification (1998): D.2, F.3, D.3, I.2, F.1, C.2

Typesetting: Camera-ready by author, data conversion by Scientific Publishing Services, Chennai, India

Printed on acid-free paper

Springer is part of Springer Science+Business Media (www.springer.com)

LNCS Transactions on Computational Science

Computational science, an emerging and increasingly vital field, is now widely recognized as an integral part of scientific and technical investigations, affecting researchers and practitioners in areas ranging from aerospace and automotive research to biochemistry, electronics, geosciences, mathematics, and physics. Computer systems research and the exploitation of applied research naturally complement each other. The increased complexity of many challenges in computational science demands the use of supercomputing, parallel processing, sophisticated algorithms, and advanced system software and architecture. It is therefore invaluable to have input by systems research experts in applied computational science research.

Transactions on Computational Science focuses on original high-quality research in the realm of computational science in parallel and distributed environments, also encompassing the underlying theoretical foundations and the applications of large-scale computation. The journal offers practitioners and researchers the opportunity to share computational techniques and solutions in this area, to identify new issues, and to shape future directions for research, and it enables industrial users to apply leading-edge, large-scale, high-performance computational methods.

In addition to addressing various research and application issues, the journal aims to present material that is validated – crucial to the application and advancement of the research conducted in academic and industrial settings. In this spirit, the journal focuses on publications that present results and computational techniques that are verifiable.

Scope

The scope of the journal includes, but is not limited to, the following computational methods and applications:

- Aeronautics and Aerospace
- Astrophysics
- Bioinformatics
- Climate and Weather Modeling
- Communication and Data Networks
- Compilers and Operating Systems
- Computer Graphics
- Computational Biology
- Computational Chemistry
- Computational Finance and Econometrics
- Computational Fluid Dynamics
- Computational Geometry

- Computational Number Theory
- Computational Physics
- Data Storage and Information Retrieval
- Data Mining and Data Warehousing
- Grid Computing
- Hardware/Software Co-design
- High-Energy Physics
- High-Performance Computing
- Numerical and Scientific Computing
- Parallel and Distributed Computing
- Reconfigurable Hardware
- Scientific Visualization
- Supercomputing
- System-on-Chip Design and Engineering

Editorial

The Transactions on Computational Science journal is part of the Springer series *Lecture Notes in Computer Science*, and is devoted to the gamut of computational science issues, from theoretical aspects to application-dependent studies and the validation of emerging technologies.

The journal focuses on original high-quality research in the realm of computational science in parallel and distributed environments, encompassing the facilitating theoretical foundations and the applications of large-scale computations and massive data processing. Practitioners and researchers share computational techniques and solutions in the area, identify new issues, and shape future directions for research, as well as enable industrial users to apply the techniques presented.

The current volume is devoted to recent advancements in autonomic computing with special focus on formal engineering methods for nature-inspired computing systems and is edited by Phan Cong-Vinh. It is comprised of seven papers providing an in-depth overview of the area and comprehensive evaluation of state-of-the-art methodologies for autonomic computing.

We would like to extend our sincere appreciation to the special issue guest editor, Phan Cong-Vinh, for his diligence in preparing this special issue. We would also like to thank all of the authors for submitting their papers to the special issue and the associate editors and referees for their valuable work. We would like to express our gratitude to the LNCS editorial staff of Springer, in particular Alfred Hofmann, Ursula Barth, and Anna Kramer, who supported us at every stage of the project.

It is our hope that the fine collection of papers presented in this issue will be a valuable resource for Transactions on Computational Science readers and will stimulate further research into the vibrant area of computational science applications.

November 2011

Marina L. Gavrilova
C.J. Kenneth Tan

Advances in Autonomic Computing: Formal Engineering Methods for Nature-Inspired Computing Systems Guest Editor's Preface

This special issue, with papers contributed by prominent researchers from academia and industry, contains reference material for researchers, scientists, professionals, and students in computer science and computer engineering as well as developers and practitioners in computing and networking systems design, providing them with state-of-the-art research findings and future opportunities and trends. These contributions include some recent advances in autonomic computing reflected in the seven papers that we chose to invite for this special issue. In particular, the special issue covers various problems on formal engineering methods for nature-inspired computing systems.

The first paper, by Arun Prakash, Zoltan Theisz, and Ranganai Chaparadza, provides a hybrid methodology consisting of formal methods to design, refine, and verify the entities of autonomic networks. The paper focuses some discussions on the methods for meta-modeling, structural modeling, and behavior modeling and design of existing protocols and newly introduced autonomic components that autonomically manage and adapt the behavior of protocols to changing policy and network conditions. A case study, based on the recently introduced Hierarchical Autonomic Management and Control Architectural Framework called GANA, is used for highlighting the practical benefits and design choices available to modelers and autonomic components designers.

The second paper, by Cem Safak Sahin, M. Umit Uyar, Stephen Gundry, and Elkin Urrea, provides a nature-inspired approach to achieving self-organization of mobile nodes over unknown terrains. In this framework, each mobile node uses a genetic algorithm as a self-distribution mechanism to decide its next speed and movement direction to obtain a uniform distribution. The paper presents a formal analysis of the effectiveness of the genetic algorithm and introduces an inhomogeneous Markov chain model to prove its convergence.

The third paper, by Phan Cong-Vinh, first constructs algebraic models for data-aware self-organizing networks (DASNs). Second, an algebraic model of data streams is developed and quantitative behaviors for data streams in DASNs are computed. Third, using algebraic models for DASNs, the paper forms monoids of data-aware self-organizations and shapes streams of data-aware self-organizations. Finally, a category of data-aware self-organization monoids is established and streams of data-aware self-organization monoids are considered.

The fourth paper, by Pruet Boonma and Junichi Suzuki, discusses inherent tradeoffs on wireless sensor networks (WSNs) among conflicting performance objectives such as data yield, data fidelity, and power consumption. In order to address this challenge, the paper proposes a biologically inspired application framework for WSNs. The proposed framework, called El Nino, models an application as a decentralized group of software agents.

The fifth paper, by Emil Vassev and Serguei A. Mokhov, discusses research towards developing special properties that introduce autonomic behavior in pattern-recognition systems. In the paper, ASSL (Autonomic System Specification Language) is used to formally develop such properties for DMARF (Distributed Modular Audio Recognition Framework).

The sixth paper, by Antonio Manzalini, Nermin Brgulja, Corrado Moiso, and Roberto Minerva, aims at looking inside the black box of an autonomic bio-inspired eco-system. Specifically, a model of autonomic component is elaborated allowing the evolution of ecosystems enabled by means of self-awareness and self-organization. This approach goes beyond a traditional mechanistic one, where concepts derived from biology are applied to explain what happens in an ecosystem.

The seventh paper, by Sylvain Halle, Roger Villemaire, Omar Cherkaoui, and Rudy Deca, stems from the observation that, when "data-aware" constraints are expressed using mathematical logic, automated sequence generation becomes a case of satisfiability solving. This approach has the advantage that, for many logical languages, existing satisfiability solvers can be used off-the-shelf. The paper surveys three logics suitable to express real-world data-aware constraints and discusses the practical implications, with respect to automated sequence generation, of some of their theoretical properties.

We owe our deepest gratitude to Dr. Nguyen Manh Hung, Rector of NTT University at Ho Chi Minh City in Vietnam for his useful support, and to all the authors for their valuable contribution to this special issue and their great efforts, and also to the referees for ensuring the high quality of the material presented here. All of them were extremely professional and cooperative. We wish to express our thanks to the Editor-in-Chief, Marina L. Gavrilova, for her important assistance with the process of assembling the special issue.

November 2011   Cong-Vinh Phan

LNCS Transactions on Computational Science – Editorial Board

Table of Contents

Formal Methods for Modeling, Refining and Verifying Autonomic Components of Computer Networks

Arun Prakash[1], Zoltán Theisz[2], and Ranganai Chaparadza[1]

[1] Fraunhofer FOKUS, Competence Center MOTION, Kaiserin-Augusta-Allee 31, 10589 Berlin, Germany
{arun.prakash,ranganai.chaparadza}@fokus.fraunhofer.de
[2] Network Management Research Lab, Ericsson Ireland Ltd., Athlone, Ireland
zoltan.a.theisz@ericsson.com

Abstract. The domain of autonomic and nature-inspired networking comes with its own set of design challenges and requirements for its architectures. This demands a tailored solution to model and design its components rather than a generic approach. In this paper, we provide a hybrid methodology consisting of formal methods to design, refine and verify the entities of autonomic networks. We focus our discussions on the methods for *meta-modeling, structural modeling* and *behavior modeling and design* of existing protocols and newly introduced autonomic components, that autonomically manage and adapt the behaviour of protocols to changing policy and network conditions. A case study, based on the recently introduced Hierarchical Autonomic Management and Control Architectural Framework called GANA, is used for highlighting the practical benefits and design choices available to modelers and autonomic components designers. The results of our case study are analyzed to explain the trade offs that future designers would be forced to make in order to achieve their design objectives for an autonomic network. A tool-chain to realize the methodology is also briefly discussed.

Keywords: Autonomics, GANA Meta-Model, Model-driven Methodology, Formal Methods, Hierarchical Controllers Design, Protocol behavior modeling.

1 Introduction

Over the past years, bio-inspired Information and Communication Technology (ICT) systems have been, and continue to be, a subject receiving continuous research. The field of Autonomic Computing and Autonomic Networking is inspired by the autonomic nervous system found in humans and other mammals. In the human body, it is understood that there are various types of systems that govern the functioning of the body as a larger entity. For example, the autonomic nervous system consists of sub-systems that are further specialized into the sympathetic nervous system ("fight or flight") and the parasympathetic nervous system ("rest and digest"). These two sub-systems are in turn co-coordinated by

M.L. Gavrilova et al. (Eds.): Trans. on Comput. Sci. XV, LNCS 7050, pp. 1–48, 2012.

the central nervous system which includes the brain and spinal cord. Thus autonomicity is involved in the development of systems that exhibit control-loop structures and mechanisms to self-manage and control the system's behavior in the face of adverse conditions and challenges to its operation, as well to changing network objectives, policies and goals that govern the operation of the system.

1.1   Motivation

1.1.1   Reflection on the Human Autonomic Nervous System

In a human body, there are different types and levels of goals that must be achieved and maintained by the body's systems in a particular context of the body's activities. Some goals are inherent to the correct functioning of the body and are handled by the sympathetic nervous system (maintains homeostasis) and its counterpart, the parasympathetic nervous system (responsible for maintaining the body at rest). Other explicit goals, like decision-making processes, are created and controlled by the brain, which also controls and maintains the autonomic nervous system and various other systems of the body. Just like in ICT systems, the concept of resources and their management also exists in the human body. Resources e.g. body organs, must also be controlled and regulated. Thus the human body exhibits control-loop structures at micro- and macro-level, i.e. at different levels of abstractions and with nesting of control-loops. The brain plays the central role at the macro-level coordinating the control-loops of the sympathetic and parasympathetic nervous systems. The autonomic nervous system's sub-systems (sympathetic and parasympathetic) work with the concept of suppression and dominance of each other in context of different body activities and experiences. Autonomic networking architectures are thus inspired from such resource management and control mechanisms.

1.1.2   Fundamentals of Autonomic Networking

The vision of the Future Internet is of a Self-Managing network. Its devices are envisioned to be designed and engineered in such a way that all the so-called traditional network management functions, defined by the FCAPS management framework (Fault-, Configuration-, Accounting-, Performance- and Security Management), as well as the fundamental network functions (such as routing, forwarding, monitoring, discovery, fault-detection, fault-removal and resilience), are made to automatically feed each other with information and knowledge (such as goals and events), in order to effect feedback processes among those diverse functions. The feedback processes enable the reaction of the various functions in the network and in individual nodes/devices, in order to achieve and maintain well defined network goals. In such an evolving environment, it is required that the network is able to detect, diagnose and repair failures, as well as adapt to changing configurations and optimize performance [1,2].

Thus, Autonomicity is an enabler for advanced self-manageability of networks, beyond what has been achieved through scripting based automation techniques. Thus, the FCAPS functions are diffused within node/device architectures, apart from being part of an overall network architecture - whereby traditionally, a distinct management plane is engineered separately from the other functional planes of the network. New concepts and functional entities with architectural design principles facilitating self-management at different levels of node/device and network functionality and abstractions are required [1,2].

Self-Management is a new and growing concept for future networks in which *Autonomic Manager* components are introduced into the node/device architectures and the overall network architecture. The *managers* are meant to *autonomically* configure, control/regulate and adapt the behavior of network protocols and mechanisms (as managed resources/entities) via the management interfaces of the individual protocols based on policy changes and dynamic *views* exposed by the managed entities to the *autonomic managers*. Such *views* can be detected incidents or state changes. Depending on their co-operative goals (discussed later), *managers* should also be able to communicate with each other through some control types of messages.

Providing a model-driven methodology to design the *autonomic managers* and their protocols and Managed Entities (MEs) is a research challenge that we address in this paper. Following the *Generic Autonomic Network Architecture* (GANA) reference model [1,2], we shall refer to the *Autonomic Manager* components as *Decision-Elements* (DEs) throughout the paper.

1.2   Structure of the Paper

The paper is structured as follows: In Section 1 the motivation for this research is introduced. Section 2 provides a background on the GANA reference model for Autonomic Networking and Self-Management and Control Theory concepts. In Section 3, the problem statement of the requirement for formal methods in autonomic network design is presented. Sections 4, 5 and 6 describe the methodology and associated formal methods to designing, refining and verifying the components of autonomic networks. A case-study showcasing the usefulness of the methodology and its methods is presented in Section 7. In Section 8, the future research directions including a tool-chain infrastructure to realize the methodology is discussed. Finally, in Section 9 the conclusions and insights of our study are provided.

2   Background

2.1   The Generic Autonomic Network Architecture (GANA)

An evolvable, standardizable and holistic Architectural Reference Model for Autonomic Networking and Self-Management within node/device and network architectures, dubbed GANA, has recently emerged [1,2]. The concept of *autonomicity* - realized through control-loop structures operating within network nodes/devices and the network as a whole, is an enabler for advanced and enriched self-manageability of network devices and networks. The central concept of GANA is that of an autonomic *Decision-Making-Element* (DME) or simply

Decision-Element (DE). A DE implements the logic that drives a control-loop over the *management interfaces* of its assigned MEs. Therefore, in GANA, self-* functionalities such as self-configuration, self-healing, self-optimization, etc, are implemented by a DE or group of DEs. The *generic nature* of GANA lies in the following properties:

- Definition of the interfaces and their fundamental operations which need to be supported by a DE,
- The interconnection and relations among the DEs within node and network architectures,
- The assignment of the MEs to DEs, including the fundamental operations that must be supported on the management-interfaces of the individual MEs,
- Specification and Description Models of the Building Blocks (DEs) and Interfaces leave out the implementation-oriented details,
- Fundamental Interfaces and Operations of the DEs must be generic to support different types of Data Models used later in the actual implementation,
- The DEs that can be instantiated by design, for a particular network device/node, are decided upon by the context and role the device/node can play in the network.

The *generic nature* is also based on the principle of separating *specifications* and *design issues* from *implementation issues*. In [2] different types of *instantiation* of GANA for autonomic network management and adaptive control of protocols in different network environments are illustrated. In [3] an instantiation case for autonomic management and control of IPv6 routing protocols and mechanisms is illustrated.

The evolvable GANA architecture [1] introduces DEs to operate at four different levels of abstraction of functionality. These DEs should be designed following the principles of *hierarchical, peering,* and *sibling* relationships among each other within a node and in the network. Moreover, these components are capable of performing autonomic control of their associated MEs, as well as co-operating with each other in driving the self-managing features of the network(s).

Figure 1 shows the generic model of an autonomic networked system and its associated control loop. This model was derived from the IBM *Monitoring-Analysis-Planning-Execution* (MAPE) model [4], specifically for addressing the autonomic networking domain. The model adopts the concept of a DE as the autonomic element, similar to the role of an Autonomic Manager in the IBM MAPE model. It also adopts the concept of an ME (Managed Entity) to entail a managed resource, or an automated task in general, instead of a Managed Element which often thought to be associated only with a Network Element (NE). In GANA, MEs are not only physical Network Elements (NEs) but also include protocol modules, mechanisms and even autonomic functionalities implemented inside a device.

The figure illustrates an important property of GANA, which is in sharp contrast to the IBM MAPE model, namely the distributed nature of the information

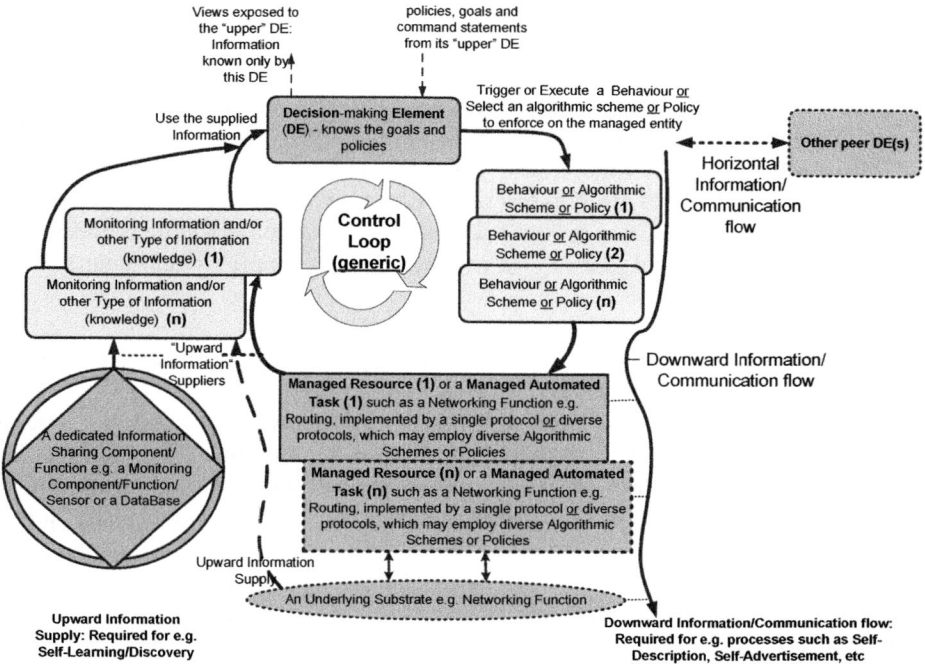

Fig. 1. A Generic Model of an Abstract Autonomic Networked System [1]

supply fed to a DE, which can be used to autonomically manage the associated MEs.

Further, the figure also illustrates the fact that the actions taken by the DE do not always necessarily involve triggering some behavior or enforcing a policy on its MEs, that is, changing its behavior. Instead, some of the actions executed by the DE may involve communication between the DE and other entities in the node/device or network architecture, e.g. other DEs in the system or in the network. This is indicated by the extended span of the arrow *Downward Information or Communication flow* and the *Horizontal Information/Communication flow* to other DEs, as well as the fact that a DE also exposes information to its upper DE and receives information, in the form of policies, goals and other commands, from its upper DE. For example, the DE may need to self-describe and self-advertise to other DEs in order to be discovered or to discover other DEs. This suggests (as shall be later discussed in more detail) that control loops in GANA can be organized into hierarchical, compound control-loop structures, in which the higher-level DEs execute management actions on lower-level DEs. This also means that an ME may be of a physical nature, or may have the nature of an automated task in abstract sense, represented by a DE, or an ME of the nature, say a protocol.

In GANA, an autonomic behavior is defined as an action of the DE in an attempt to regulate the behavior of the MEs associated with it. The autonomic behavior is either spontaneously started by the DE, or it can be triggered upon receipt of information from the information suppliers of the DE, i.e., its associated MEs or sibling, peer or upper DEs. Such information can be events relevant to the state of the system, security threats, incidents, etc. A behavior triggered spontaneously by a DE, on the other hand, is simply a spontaneous transition in the finite state machine describing the overall behavior of the DE. Examples of autonomic behaviors are self-description, self-advertisement, self-healing, and self-configuration, all triggered by a DE. An autonomic behavior, therefore, always binds to a DE and possibly to the information supplier components of the control-loop.

2.1.1 The GANA Levels - Autonomicity and Self-management in GANA

GANA fixes four basic hierarchical levels of abstractions for which the generic model of an autonomic networked system presented above, i.e. DEs, MEs, Control-Loops and their associated dynamic adaptive behaviors can be designed, capturing a *holistic view of inter-working autonomic and self-management levels*. The levels of abstractions are as follows:

Level-1: Protocol-Level (the lowest level) in which self-management is associated within a network protocol (whether monolithic or modular). However, there is a growing opinion that future protocols need to be simpler, with no decision logic embedded, than today's protocols which have become too hard to manage. This means that there is a need to rather implement decision logic at a level higher, i.e. outside the individual protocols [3,5,6].

Level-2: The abstracted function-level is directly above the *protocol-level*. It abstracts some protocols and mechanisms associated with a particular function such as *routing function, forwarding function, mobility management function'*, etc. At this level we can then reason about autonomic routing (see example instantiation of GANA for realizing autonomic routing in [30]), autonomic forwarding, autonomic fault-management, autonomic configuration management. etc.

Level-3: The level of the node/device's overall functionality and behaviour i.e. a node or system as a whole is also considered as a level of self-management functionality. At this level of self-management, the DEs operating on the level of abstracted networking functions become the MEs of the Main DE of the node. This means the DEs at this level have access to the *views* exposed by the lower level DEs. The Node-Main-DE uses this *knowledge* to influence the lower level DEs to take certain desired decisions. This in turn may further influence or enforce desired behaviors on their associated MEs, inductively down to the lowest level of individual protocol behavior.

Level-4: This is the highest level in the GANA hierarchy and abstracts the overall functionality and behavior of the network. Network-Level DEs are characterized by the following properties:

1. they have broader network-wide *views* to perform sophisticated decisions e.g. network optimization,
2. they are logically centralized to either avoid processing overhead in managed nodes or scalability and/or complexity problems with implementing distributed decision logic in network elements,
3. they are the ones that provide an interface for a humans to define Goals and Objectives or Policies e.g. Business Goals.

Figure 2 illustrates that, at node level of self-management, the lower level DEs operating at the level of abstracted networking functions become the MEs of the main DE of the node. This means the node's main DE has access to the *views* exposed by the lower level DE and uses this overall knowledge to influence and enforce the lower level DE to take certain desired decisions. This may in turn further influence or enforce desired behaviors on their associated MEs, down to the lowest level of individual protocol behavior. A *Sibling* relationship simply means that the entities are created or managed by the same upper level DE.

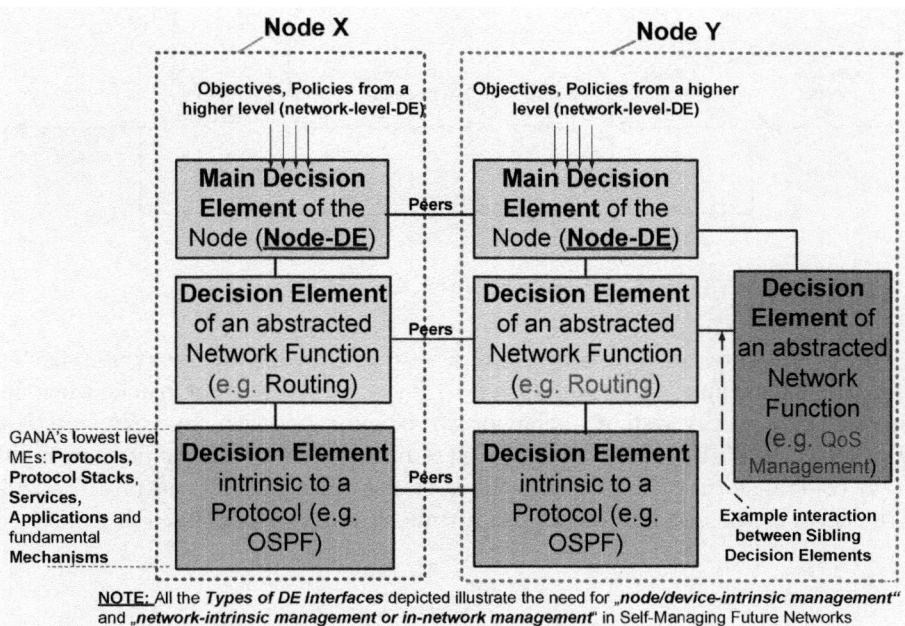

Fig. 2. Illustration of Hierarchical, Peering, Sibling Relations and Interfaces of DEs in GANA that require to be specified [1]

This means that the entities having a sibling relation can form other types of peer relationship within the autonomic node and with other DEs hosted by other nodes in the network. Thus in GANA:

- Lower level DEs expose *views* up the Decision Plane, allowing the upper (slower) control loops to control the lower level (faster) control-loops.
- The decisions computed in the upper DEs implementing slower Control-Loops are propagated along the DE hierarchy to the Functions-Level DE(s) implementing the faster control-loops, which then arbitrate and enforce the changes to the lowest level MEs (protocols and mechanisms). For more detailed descriptions of GANA, we refer the reader to [1,2].

2.2   Modern Control Theory

Control Theory is the science of controlling a system behavior by manipulating the inputs of the system under control. A simple feedback control system is portrayed in Figure 3. A feedback control system is defined as one that evaluates the difference between the measured output (feedback) and the reference input in order to calculate the control input to regulate the system under control [7,8].

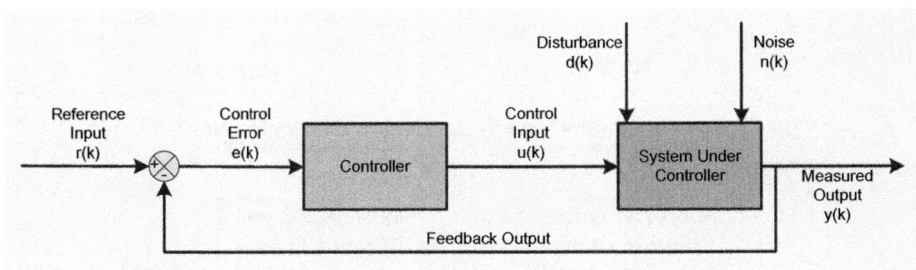

Fig. 3. A simple Feedback Control System [7]

As we are interested in controlling a system's (protocol) behavior reflected through its outputs, it is imperative to know how a system reacts when its inputs are varied. A system's behavior can be expressed as a *Transfer Function*. As defined in [7] a *transfer function* $G(z)$ of a system describes how an input $U(z)$ (Z-Transform of the input $u(k)$) is transformed into the output $Y(z)$ (Z-Transform of the output $y(k)$). Thus a transfer function is defined as

$$G(z) = \frac{Y(z)}{U(z)} \tag{1}$$

The reaction of a system's behavior to the variation of its inputs can be understood by stimulating the system with a step input. A step input $(U_{step}(k))$ is defined as follows:

$$U_{step}(k) = \begin{cases} 1 & \forall k \geq 0 \\ 0 & otherwise \end{cases} \tag{2}$$

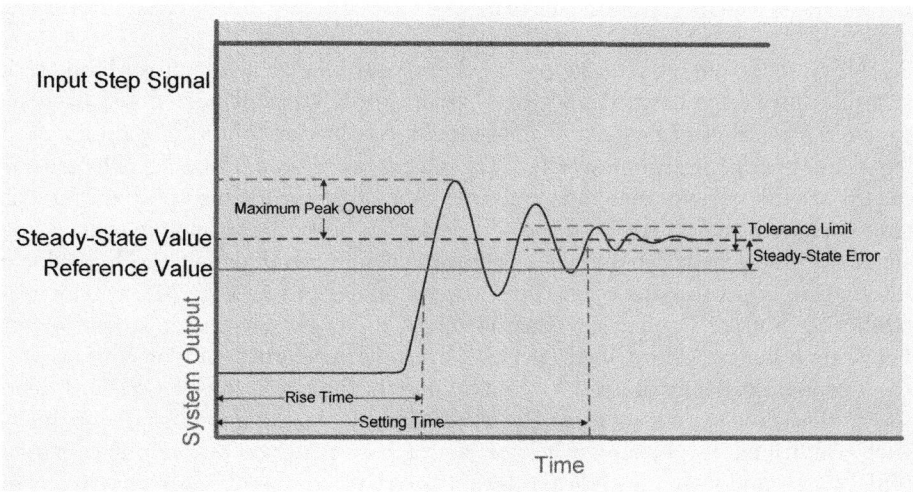

Fig. 4. Transient Response of a Theoretical System's Behavior

A theoretical system's response (transient response) to a step input is shown in Figure 4. *Transient Response* is defined as the system's behavior before it reaches the steady-state value of the behavior. *Steady-State Value* is defined as the output of the behavior at infinite time, if the output converges. We are interested in some of the properties of this transient response as they provide valid insights into the inner workings of the system under control. These properties determine how a controller is designed to manage the system. For computing systems, we are interested in four specific properties, defined as *SASO* properties. Based on the definitions provided in [7,8] we redefine these properties with modifications for the autonomic networking domain as follows:

- **S***tability:* A protocol's behavior is defined as BIBO (Bounded-Input-Bounded-Output) stable, if and only if all the poles of the closed-loop transfer function representing the behavior of the system and the controller are inside the unit circle of the complex plane. There are different tests (criteria) for checking the stability of a system if the behavior model of the system and its controller are expressed as State-Space Models. State-space models represent the system dynamics through auxiliary variables called state variables in addition to input and output variables. They are an alternative to the transfer function models used in classical control theory.
- **A***ccuracy* or *Steady-State Error* can be defined as the difference between the steady-state value and the expected (wished) reference value of a behavior output.
- **S***ettling Time* is defined as the maximum time taken by the protocol's or ME's behavior to reach its steady-state value.
- *Maximum Peak* **O***vershoot* is defined as the maximum amount by which the transient response exceeds the steady-state value of the output.

3   The Problem Statement

As addressed in [9], control-loops in general and hierarchical control-loop structures of autonomic networks need to be cautiously conceptualized and designed to avoid a number of problems. These include behavior conflicts, stability problems [10], operating regions overlap [11] and response time problems. The authors of [12] provide a taxonomic and tangible view of the problems and requirements that designers should consider for addressing stability in autonomic network designs. Conflicts in the behavior of a protocol may occur when the behavior of interacting control-loops operating over the same protocol or ME render contradicting and/or conflicting control-inputs to it. For instance, an autonomic behavior managing the timers of OSPFv3 may contradict and/or conflict with the autonomic behavior (self-*) controlling the link weights of OSPFv3, leading to disastrous consequences. Response time problems may also occur during such conditions. In [3], the authors present a case study on autonomic control of OSPFv3 behavior at multiple levels that need to co-operate so as not to introduce conflicting behavior. Even though the design principles of some autonomic architectures such as GANA partially contribute to address the above mentioned problems, a formalized methodology is imperative for the design and verification of hierarchical controllers of any autonomic network. By verification we mean that the designed autonomic entities/hierarchical controllers can be verified to be free from the above mentioned problems, and if detected in the system model, resolutions need to be provided.

The importance of Formal Description Techniques (FDTs) in systems design is well known. A model-driven approach to DE design, for addressing potential problems of stability in control-loops, while still at design time, is required. Modeling and validation of autonomic behaviors of DEs using FDTs, such as the well-known and successful ITU-T SDL [13] language would thus need to be explored. Such an approach would enable the design, model-checking, verification some partial code-generation of DEs, and the validation and simulation of autonomic behaviors. The aspects of stability of control-loops that need to be addressed at design time are:

1. Requirement for a GANA Meta-Model that enforces constraints in the design models for DEs and their interactions with assigned MEs and with other DEs, that should be embedded in the Modeling Tools e.g. a *model editor*, is required,
2. A Model-Walker can be designed for walking over a design model in order to detect conflicting control-loops and overlaps in the so-called *valid operating regions* [7,11,8] of control-loops specific to DEs.
3. Simulations for detecting behavior conflicts between interacting control-loops designed to operate within a node/device and in the network (outer loops).

Providing such a formalized model-driven methodology and tool-chain is a research challenge. In this paper we address this challenge by providing a methodology composed of formal methods to model, design, refine and verify autonomic components of computer networks.

4 Methodology

A methodology with formal methods to design, verify, validate and simulate autonomic entities is a crucial prerequisite for the proper functioning of an autonomic network with hierarchical controllers. The authors of [9] have captured the requirements, which every methodology that wishes to provide a formalized approach to design and verify autonomic entities should realize and satisfy. In [14], the authors have provided a methodology to design hierarchical controllers for network resource management. Their methodology provides excellent dependability evaluation for the design of hierarchical controllers. However, in the context of autonomics, we believe that the methodology described as such, is deficient, as we believe that autonomic entities are not limited in their functionality and role as controllers. Instead, they can be seen as advanced decision making elements that perform other self-* tasks, like self-configuration, auto-discovery to name a few, in addition to controlling the underlying MEs through their controllers.

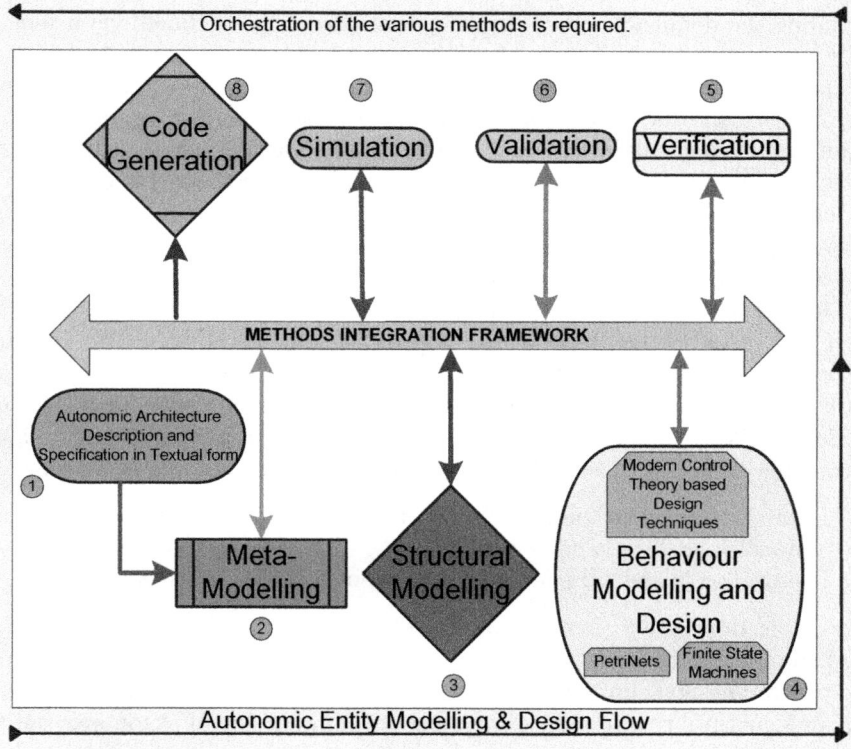

Fig. 5. Model-Driven Methodology with formal methods to Design, Verify, Validate and Simulate Autonomic Components of a computer network

With this in mind, we provide a model-driven methodology, that satisfies the structural, behavioral and stability requirements documented in [9]. The methodology in its holistic form is shown in Figure 5. As it can be seen, the methodology is composed of the following methods: *Meta-Modeling, Structural Modeling, Behavioral Modeling and Design, Verification, Validation, Simulation* and *Code-Generation.*

The *Meta-Modeling* method furnishes a versatile state of the art software engineering approach that relies on well-proven domain modeling techniques, standardized meta-modeling languages such as OMG's MOF [15] and flexible tool support ranging from model editors and validating engines to model-to-model translators and intelligent template generators. In general, the potential benefits of Model Driven Architecture (MDA) [15] and software design are well understood and unequivocally demonstrated within the contemporary software industry. However, its introduction into other technical domains is still considered to be a challenging endeavor. Furthermore, it goes beyond the widely accepted limits of UML [16] modeling.

In the frame of autonomics, a more formal requirement specification of an autonomic architecture could benefit enormously, if its particular technical domains, including the details of some relevant particularities are defined, via a semantically precise meta-model ensemble. The meta-model should aim to deal with and formalize as much as practically possible, each and every aspect of an autonomic architecture and its mechanisms. This would assist in analyzing, modeling, designing, verifying, validating and building autonomic components through semi-automated means that are compliment with their autonomic architectures. Thus we have a single *Meta-Model* consisting of several interconnected sub meta-models showcasing the concepts and the structural and behavioural relationships among the various meta-entities of an autonomic network architecture.

The methods, *Structural Modeling* and *Behavioral Modeling and Design* base their processes on the assets of the *Meta-Modeling* method, namely the single *Meta-Model*. Unlike conventional protocol design techniques for computer networks, where the behavior of protocols are described as Finite State Machines (FSMs) or Petri-Nets (PNs) and documented in RFCs, modeling and designing autonomic entities require a different approach due to a couple of reasons:

1. There is a need to enforce an autonomic architecture on top of existing protocols and MEs, and
2. The protocols and MEs already have their behavior specified (in a RFC).

Thus it is imperative that the methodology enforces an orchestration of the methods *Structural Modeling* and *Behavioral Modeling and Design* in the specified order. The *Structural Modeling* method ensures that an autonomic architectures's structural requirements, in terms of the hierarchy of autonomic entities, their relationships with each other, their interfaces, their data types (both internal and external), communication channels, and so on are specified before the behavior of each of these autonomic entities are designed or modeled. On the other hand, the *Behavioral Modeling and Design* method would focus on modeling the behavior of existing protocols (or MEs) and designing the behavior of the entities that manage these protocols. It will be based on the structure that was specified for these protocols and autonomic entities. Even though a strict orchestration of these methods is enforced, the methodology allows the refinement of the structural models once the behavioral models have been specified. This flexibility (or iteration) is necessary as some of the structural features required for the full expression of an entity's behavior can be missed out by a designer, during the initial structural modeling iteration. As both *Structural Modeling* and *Behavioral Modeling and Design* methods are based on the same meta-model, the refinement of the individual models does not do break the model continuum or the meta-model, and is always in sync.

The backbone of the methodology is the *Methods Integration Framework.* This framework ensures seamless integration, orchestration and sharing of the modeling and data assets of the various methods. While most system engineering methodologies [17] are strictly linear and/or iterative, with their apparent strengths and weaknesses, the methodology described here is rather only loosely linear and iterative. Its a broadly decentralized methodology where all the methods share a single model perspective and operate almost simultaneously and independently of each other. It is the *Methods Integration Framework* that allows for this near simultaneous and independent model development, verification, validation and simulation of autonomic entities. The model, its verification, validation and simulation results, data and insights can be shared by all the methods, enabling model (and if necessary, meta-model) refinements. Finally, this methodology is aptly suited to be realized as a tool-chain, allowing individual tools realizing the respective methods to be integrated and orchestrated through an integration tooling infrastructure.

In the sections below, we restrict our discussion to the details of the *Meta-Modeling, Structural Modeling* and *Behavioral Modeling and Design* methods. A detailed discussion of the other methods is beyond the scope of this paper.

5   Meta-Modeling and Structural Modeling of an Autonomic Architecture and Network

Modeling the architectural, logical and functional structure of an autonomic architecture, such as GANA, provides a formal representation for its constituent elements such as DEs and MEs and firmly establishes their modus operandi. The rules, as to how the individual control-loops are imagined, and permitted to interact and influence each other's functionality is thus defined. Obviously, the models of the various GANA incarnations must follow certain design concepts. Hence, at the core of the modeling efforts lays the GANA meta-model, which provides an ideal platform to specify the structural, logical and functional requirements of any GANA compliant autonomic network.

In general, the GANA meta-model, that is explained in detail in the sequel, fully utilizes the facilities of a meta-modeling approach - the Model Integrated

Computing (MIC) [18,19,20,21]. It therefore formalizes all relevant and distinguishing aspects of the GANA architecture and its incorporated autonomic mechanisms. This assists in the analysis, verification and final evaluation of GANA compliant use case scenarios by semi-automated means and fashion.

The syntax and static semantics, depicted in this paper, of the GANA meta-model are expressed in a MIC compatible visual meta-modeling language, based on UML class diagram notation and accompanying OCL constraint description, visualized via the language default editor, Generic Modeling Environment (GME) [22]. Further, for the purpose of clarity and focus, the meta-models described in the paper have been deliberately deprived of their attributes. Thus the focus of the paper is on expressing the concepts and the relationships among the various meta-entities by establishing their complete intentional semantics. For the complete meta-model, including the detailed representation of the attributes, the reader is referred to [23].

Although OMG's MOF is the industry's de facto standard of meta-model design, our selection of GME provides some extra meta-modeling facilities. The GANA meta-model takes advantage of these in order to create a meta-modeling environment that support seamless distributed re-use of autonomic sub-patterns. This technique is not, by any means, a unique property of MIC. However, its representation is much more straight-forward in GME than in any MOF compatible editors. Therefore, it benefits the reader to fully comprehend the novelty of this concept of GANA meta-modeling.

5.1   Elements of GANA Meta-Model

5.1.1   Control-Loop

One of the major aims of the GANA reference model is to amend the shortcoming of past and recent autonomics related approaches when it comes to establishing evolvable architectures for the self-managing future networks. Previous to GANA there have not been any serious attempts to holistically specify the control-loops required for autonomicity, at both micro- and macro-level. This includes their interactions and relationships, for nodes and the network as a whole, got by explicitly capturing the diversity in information suppliers that feed and drive the control-loops. In general, the concepts of a *Control-Loop, Decision-Element, Managed Entity* or *Managed Automated Task*, as well as the related self-manageability issues, may be associated with some implementation of a single network protocol. They can be implemented within the holistic behavior of the protocol, say, OSPF.

Figure 6 shows an example of a protocol-intrinsic control loop with a DE, which constitutes the core logic of the protocol, driving a control-loop that is intrinsic to the protocol and thus referred to as a Protocol-Level DE in this case. Although protocol intrinsic is the implementation level of the current Internet protocols, we firmly believe that in next generation networking, these MEs will be refined. Therefore, control-loops are one of the most important concepts of the GANA meta-model.

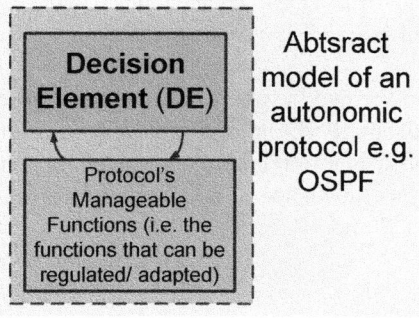

Fig. 6. A Protocol-Intrinsic Control-Loop - Protocol-Level Decision-Element

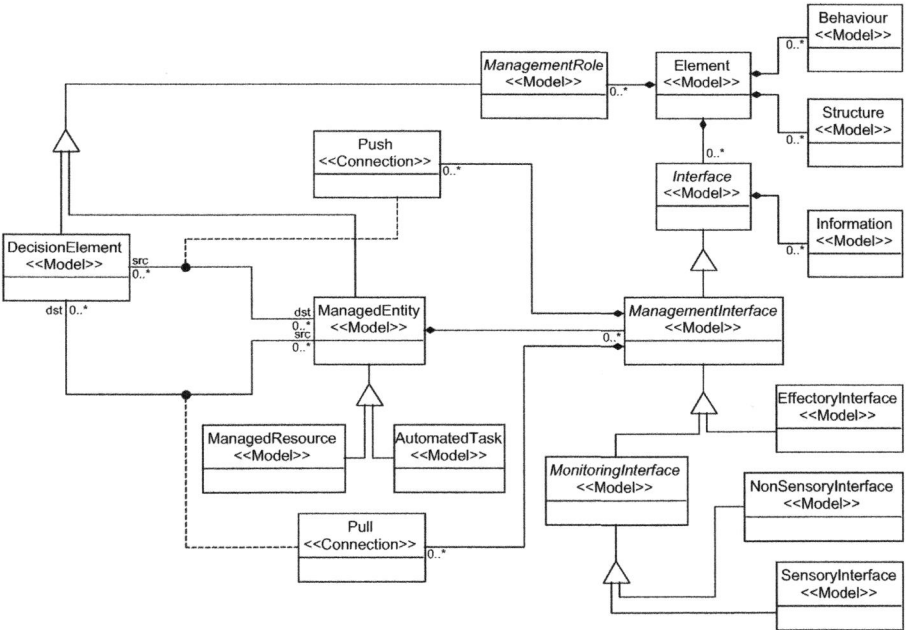

Fig. 7. Control-Loop Internals

The meta-model snippet in Figure 7 depicts the constituent elements of a control-loop. A control-loop is defined by a DE which acts upon an ME. The GANA reference model enables hierarchical control-loops at three different levels, with different timescales and control behaviors. Hence, in general, a DE of a particular control-loop can play, at the same time, the role of a ME in another control-loop. Consequently, an abstract <<Model>> *Element* is introduced into the meta-model of the control-loop. This abstract *Element* represents the core part of the control-loop. Both the structural decomposition and behavior description of a control-loop are harmonized with the requirements of the GANA reference model.

The communication facility between *Elements* is explicitly specified via management interfaces, although the generic interface structure is identical in the case of the DE and ME. The particular operations they provide are specific according to their role, as shown for a DE in Figure 8, and for an ME with two alternatives in Figures 9 and 10.

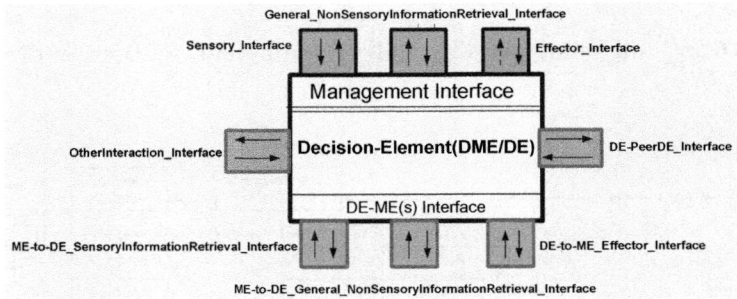

Fig. 8. Interfaces of a Decision-Element

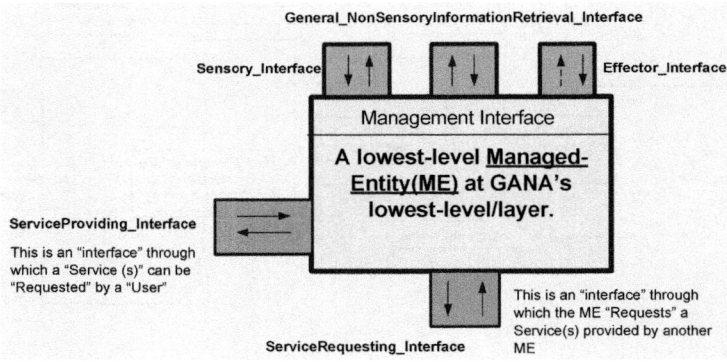

Fig. 9. A Managed Entity at GANAs lowest layer - Variant A

Taking this into account, a detailed interface compliance should be safeguarded by specialized tools within the tool-chain. The meta-model only prescribes that interfaces must be provided and should be according to the role the *Element* is required to play in the control-loop. The interaction between the DE and the ME consists of mechanisms and protocols that implement a

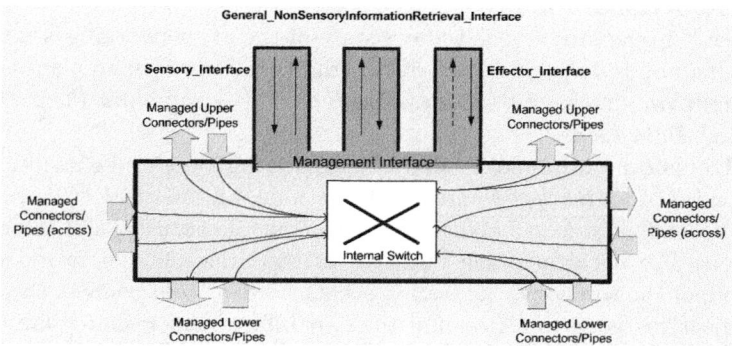

Fig. 10. A Managed Entity that is an Evolved Protocol or a Future Protocol Model in GANA - Variant B

robust and efficient communication substrate for exchanging control information as well as any special type of knowledge via push or pull models for information retrieval. Both of these capabilities are cared for in the meta-model via the explicit <<Connection>> *Push* and *Pull*, respectively.

Due to the hierarchical and distributed nature of control-loop establishment in GANA, the meta-model must also take into account the means for DE-to-DE and DE-to-ME interactions. As shown in Figure 11, the different interaction modes are modeled via <<ConnectionProxy>>. This is an advanced MIC feature to reference already defined elements of the meta-model. This ensures that the meta-model is modular, re-usable and compliant with the design guideline for the separation of concerns. These interactions are vital, and important to be precisely modeled in any use case scenarios. Only via this explicit entanglement, is it possible to methodologically evaluate the viability of the control-loops against the required stability, performance and scalability constraints.

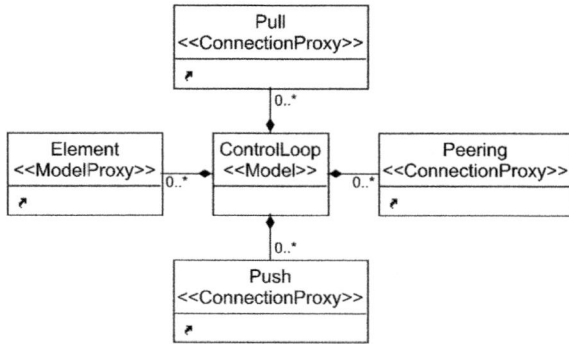

Fig. 11. Meta-Model of Control-Loop Entanglement

5.1.2   Functional Planes

In the GANA reference model, today's state-of-the-art networking is compressed, by merging and re-factoring some of the traditional networking planes, into four functional planes referred to as *Decision Plane, Dissemination Plane, Discovery Plane* and *Data Plane*.

The **Decision Plane** makes all decisions driving a node's behavior, including the behavior of all managed entities of the node, and network-wide control, including reachability, load balancing, access control, security, and interface configuration. The decision plane operates in real time on a network-wide view consisting of the topology, the traffic, events, context and context changes, network objectives/goals/policies, and the capabilities and resource limitations of the nodes and devices of a network of some scope. The elements of the decision plane are considered as autonomic elements, that is, DEs. These DEs drive the control-loops of an autonomic node/network forming a hierarchy that consists of four levels of autonomicity. The corresponding snippet of the meta-model is shown in Figure 12.

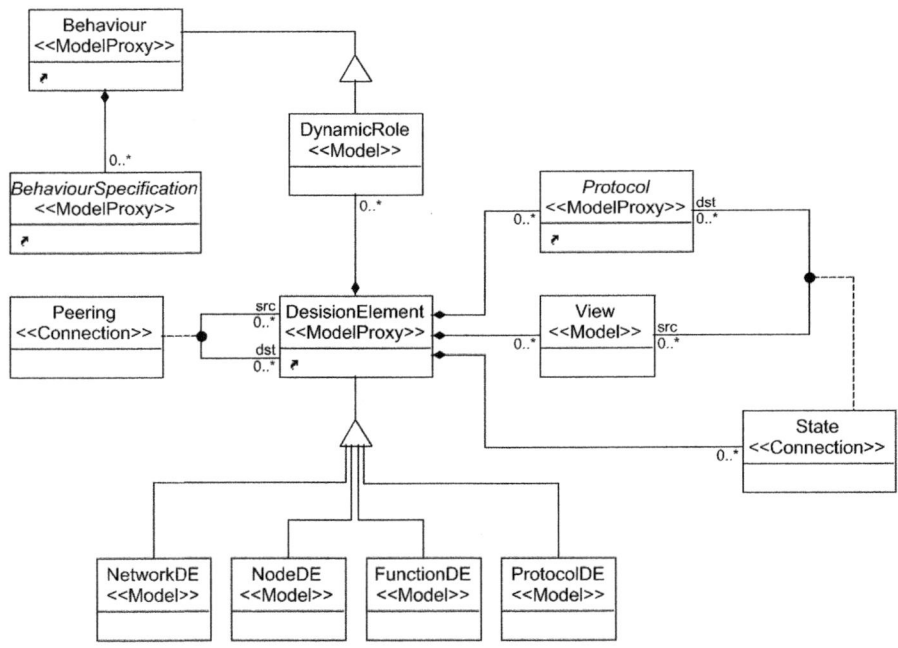

Fig. 12. Meta-Model of GANA Decision Plane

Firstly, the DEs are categorized into four subtypes (see levels of GANA in Section 2.1.1) in line with their scope of responsibility. Secondly, any interaction that may operate between the DEs must to be based on some real protocol. However, not all details of that protocol can or should be taken into account when it comes to focusing on the control related aspects of the facility. Therefore, DEs, and also MEs, only view and expose part of their internal mechanism via some control-related specific views or through observable impacts, to the autonomic mechanism provided by the DE. The parts of, e.g. the internal protocol, that do not contribute directly, are abstracted away.

Finally, DEs possess dynamic roles - at least one, but more compatible overlapping roles are also allowed - that specify the behavior of that particular element according to the responsibility it has in a particular control-loop. This role is associated with a certain behavior specification that represents non-structural aspects, such as FSM like behavior, which, in such a manner, describes the dynamicity of the DE specific control policy it contributes to.

The **Dissemination Plane** consists of mechanisms and protocols that provide a robust and efficient communication substrate. This is used to exchange control information as well as any special type of information, or knowledge, that is not considered as the actual user data, e.g. monitoring data, among DEs inside a single node, and among nodes. The *Dissemination Plane* can be used for conveying the following types of information or knowledge:

- Signaling information,
- Monitoring Data, including change of *State Info*,
- Other types of Control information that need to be exchanged between DEs,
- *Incidents Info* e.g. faults, errors, failures, alarms, etc.

Example elements of the *Dissemination Plane* that can be considered to belong to this plane are protocols or mechanisms such as: ICMPv6, MLD, DHCPv6, SNMP, IPFIX, NetFlow, IPC mechanisms. The meta-model in Figure 13 shows only a subset of the potential types of information the *Dissemination Plane* can carry. Of course, the meta-model can be extended on a case by case approach. In practical terms, any non-monotonic modification of the meta-model is automatically carried over into the next version of compatible scenario models by GME.

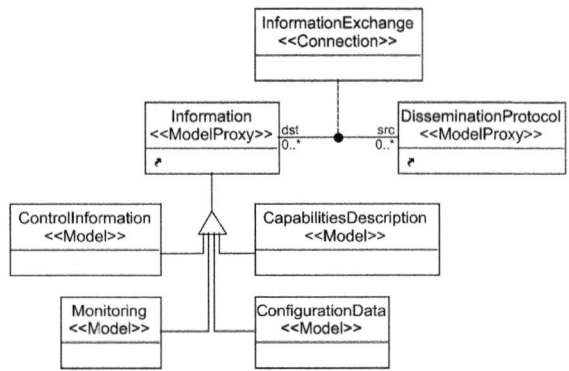

Fig. 13. Meta-Model of GANA Dissemination Plane

The **Discovery Plane** consists of protocols or mechanisms responsible for discovering entities that make up the network or a service and for creating logical identities to represent these entities. The *Discovery Plane* defines the scope and persistence of the identities, and carries out the automatic discovery and management of the relationships between them. This includes node-level discovery, neighbor-discovery, including the *Capabilities* of nodes of interest e.g. routers that are members of the all-routers multicast address of which this node is a member, network discovery, service-discovery, etc. Protocols and mechanisms for self-description and self-advertisement of *Capabilities* of entities are at the heart of this plane with example elements such as IPv6 Neighbor Discovery, IPv6 SEND, Service-Discovery Protocols/mechanisms, Topology-Discovery Protocols, and more.

The **Data Plane** consists of protocols and mechanisms that handle individual packets, up to the traditional layer 4 protocols of e.g. TCP or UDP, based on the state that is output by the *Decision Plane*. This state includes the forwarding tables, packet filters, link-scheduling weights, and queue management parameters, as well as tunnels and network address translation mappings.

Similar to the case of the *Dissemination Plane*, but maybe even more dominantly, the protocol specific aspects of the *Discovery* and the *Data Plane* cannot be foreseen once and for all. Therefore, the meta-model must show flexibility in regards of extendibility. So that the part of the meta-model, depicted in Figure 14, is, by nature, incomplete, though it shows explicitly the most important concepts of the *Discovery Plane*.

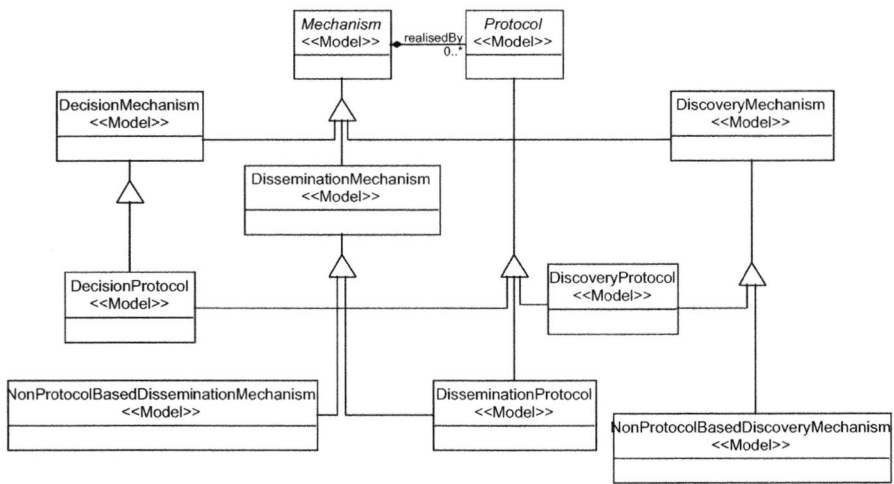

Fig. 14. Meta-Model of Protocols

5.1.3 Dynamic Behavior

The GANA reference model has been conceived to support a highly distributed autonomic control system in the domain of next generation Internet protocols.

Being a protocol aware and networking enabled control architecture, the atomic constituents of its dynamic behavior descriptors can be formally represented within the meta-model. Taking into account that GANA fully supports the evolutionary approach in the migration towards Future Internet, it is almost obvious to see that the basic building blocks of the dynamic behavior descriptions are: *Events, Actions* and *Conditions*. Of course, GANA does not want to constrain, in any practical sense, the semantics of the technique used to specify the dynamic behavior of a DE. Hence, Finite State Machines (FSM) are as well supported, among others, as e.g. Hierarchical Colored Petri Nets. This liberal support of different semantic descriptors might jeopardize the total consistency of a complex use case scenario model where one part of the control system is given via FSM, the other part is specified via policies etc. Therefore, the GANA meta-model takes advantage of a particular MIC feature to safeguard that high level of model sharing and corresponding blueprint reuse will be available.

The idea can be explained best in the case of the events. In a complex system, either a new event is introduced or an existent event is reused. Here reuse is meant on the model level and not on the instance level at run-time. In other words, the meta-model must not differentiate whether a <<Model>> *Event* or just a <<Reference>> *Event* is applied in a state of an FSM. In practical term, it means that the *Event* is defined fully only at one single place and then only extended and reused. The MIC meta-modeling pattern that makes it possible is shown in Figure 15. By the introduction of First Class Object, <<FCO>>, the event embedding part of the meta-model does not know, and, obviously, does not want to know the internal structure of the event if it does not explicitly rely on a particular feature thereof.

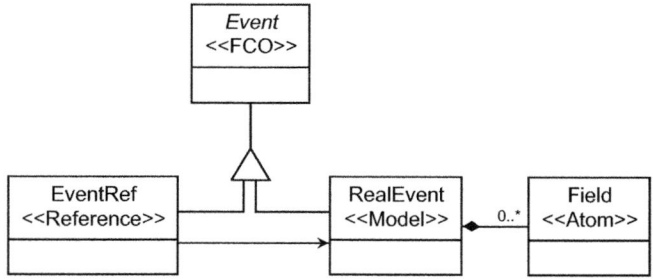

Fig. 15. Meta-Model of Events

Having this meta-modeling pattern explained, let us put it into the perspective of the GANA reference model. It is very probable that with the Future Internet only the protocols will be standardized and not the actual control semantics. Therefore, although the events are being shared between two interacting DE peers, it might happen that they operate following different principles; one driven by an FSM, the other controlled by a policy based mechanism. Nevertheless, they want to share as many concepts as possible such as guard conditions and actions whenever a state is entered or a rule is fired. Or, one wants to extend the action(s) of another already defined element or vice versa. Moreover, the whole system consists of many DEs that borrow events, actions and conditions from each other and can freely modify, extend or reuse them for serving their own purposes in a full-scale ecosystem of co-operating control nodes. Fortunately, the pattern depicted in Figure 15 makes this *model-patent* sharing functional if it is applied for all the three main types of atomic concepts, that is, for *Events*, *Actions* and *Conditions*.

5.1.4   Events and Actions

The generic control-loop dynamicity necessitates, as a bare minimum, some sensory input and effectory output. The meta-model provides these elements in the form of event hierarchies and action workflows. Moreover, the availability of Network-Level control-loops in the GANA reference model further extends this simple modeling assumption by explicitly requiring that, wherever an event or action concept fits in on its own right a reference to it must fit in too.

In practical terms, it means that events or arbitrary parts of any action workflows can be either newly introduced by a certain DE, or an already introduced event or action can be reused via an explicit reference pointing to it. Figure 15 and Figure 16 depict the *Event* and *Action* workflow meta-model snippets, respectively. The event classification is kept here to the bare minimum because the protocol related 'real' events simply create a protocol dependent inheritance hierarchy, and thus must be developed on a case by case basis. Of course, if protocols share common functionality, by extension, versioning or practical reuse, the corresponding events are reused via their references, totally interchangeably with the originals, during model building.

The *Action* meta-model snippet, shown in Figure 16, is also very generic. It enables the establishment of various kinds of expressive, federated workflows built over the atomic action types of communication, computation, direct effect and their arbitrary combination by parallel, sequential and conditional constructs. Although there are many well-known de facto standardized workflow languages, e.g. BPEL, the extra referencing capability introduced by this workflow language provides a niche advantage when it comes to specifying network level control-loops.

Both events and actions can be triggered or limited in their functionality by explicit guard conditions. Therefore, the meta-model incorporates also the concepts of the canonical representation of logical expressions, in the same *model-patent* reusable mode, as it is shown in Figure 17.

5.1.5   Behavior Paradigms

The GANA reference model is essentially protocol driven. Therefore, any compliant use case scenarios and instantiations of the GANA meta-model, are best

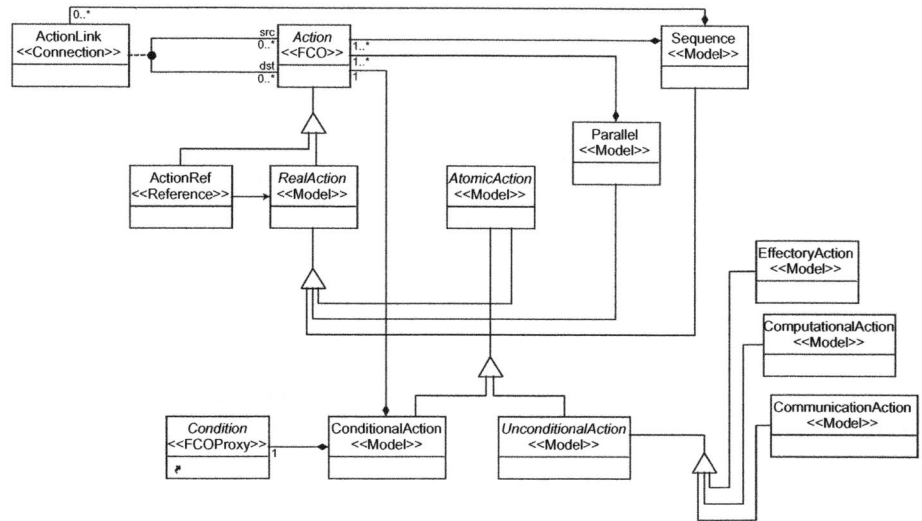

Fig. 16. Meta-Model of Actions

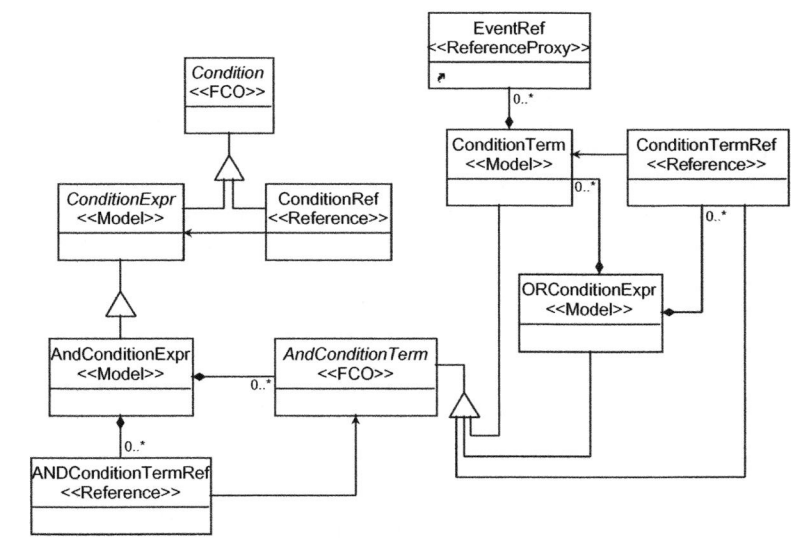

Fig. 17. Meta-Model of Logical Expressions

specified via dynamic semantics description formalisms. There are many a galore of well researched such formalisms in the scientific literature, i.e. *FSM, CFSM, Colored Petri Net, State-Space Models, First Principles Models* just to mention the most applied ones. We intended to enrich the GANA meta-model with as many alternatives as possible, taking into account some engineering trade-offs. The current meta-model is depicted in Figure 18.

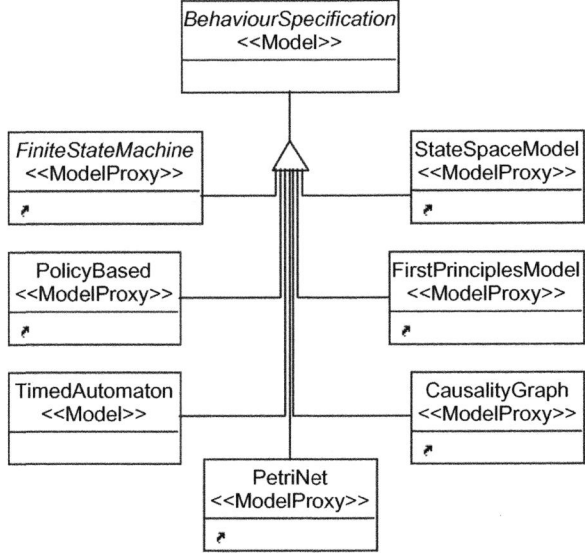

Fig. 18. Meta-Model of Behavior Paradigms

5.1.6   Policy Based Control

The policy based control paradigm, shown in Figure 19, is the most straightforward way to combine the elements of the dynamic part of the GANA metamodel, i.e. *Events*, *Conditions* and *Actions*, according to the Event-Condition-Action principle of any policy based rule system. Moreover, the meta-model further extends this trivial policy based principle realization by allowing arbitrary workflows in place of atomic actions, making the design suitable to specify complex rule dynamics of non trivial autonomic use case scenarios.

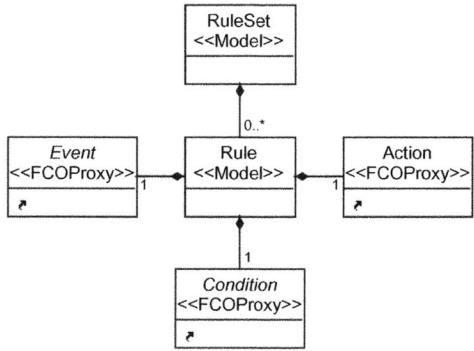

Fig. 19. Meta-Model of Policy Based Control

5.1.7   Semantic Arc

There are situations in complex autonomic control exercises, where the exact nature of a dynamic behavior is either impossible to discover or unnecessary, in practical terms, to be specified in fine details. However, causality relations between events and actions must be explicitly represented. Fault management is an almost perfect domain to showcase these situations, as in many practical cases, the root cause analysis is carried out only via statistical means. In those circumstances the probabilistic dependency graphs provide the best effective representation of any causality concerns. The corresponding part of the meta-model is presented Figure 20.

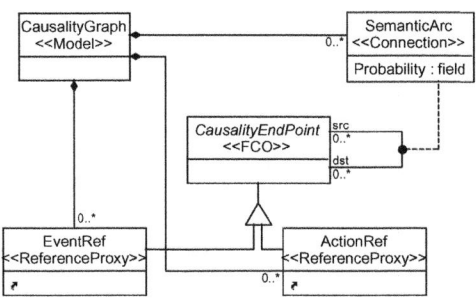

Fig. 20. Meta-Model of Semantic Arc

5.1.8   Finite State Machine and Colored Petri Net

The principles of both FSMs and Colored Petri Nets (CPNs) rely heavily on the idea of explicit state and transition representation. This idea is the "common denominator" within the range of similar paradigms. Thus it has been meta-modeled separately, as shown in Figure 21, in accordance to the generic modularity principle of separation of concerns.

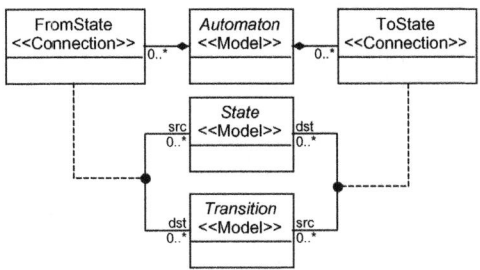

Fig. 21. Meta-Model of Explicit State Transition

Capitalizing on explicit state representation, further behavior specifications give way to two relatively different specification techniques. One for orchestration and one for choreography. In the case of orchestration, the point of view is mainly focused on the internal behavior representation of a particular DE, without taking into consideration the exact structure of the environment it is operating in. Hence, the behavior specifications of DEs are locally scoped, reliant only on direct triggering originating from the local environment, and specifies merely the local consequences induced by the DEs. Nevertheless, thanks to the extra referencing facilities to events and actions, the GANA meta-model makes it possible that any local behavior specification could refer to any action that has been defined in a totally different part of a particular scenario model. The two types of orchestration means for specifying the dynamic behavior of higher level autonomic control are the well known Mealy and the Moore automata depicted in Figure 22.

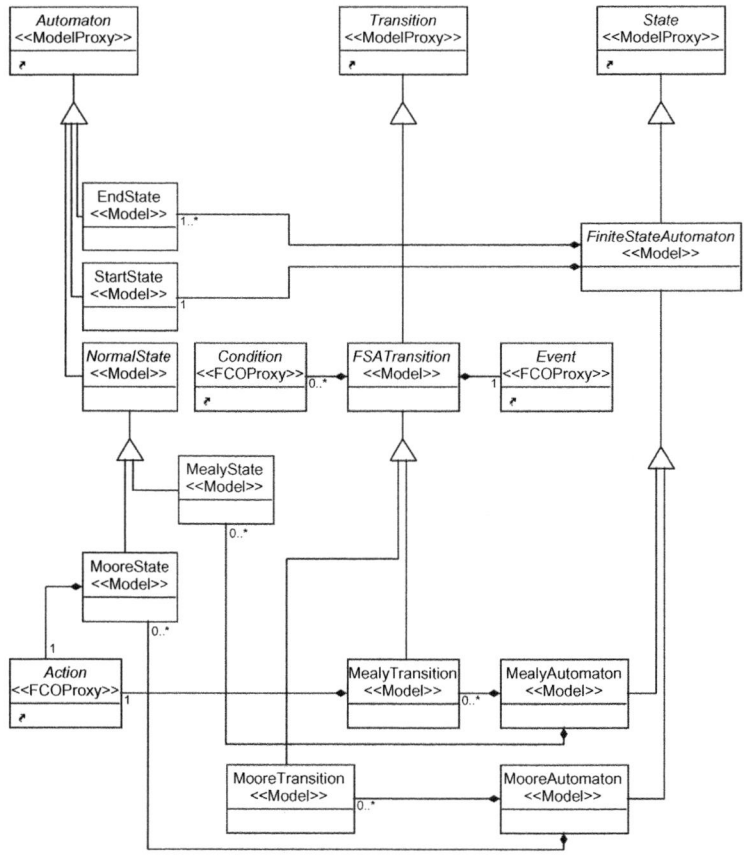

Fig. 22. Meta-Model of Finite State Machine

Choreography can play an important role in the behavior specification of network level GANA control-loops. It is mainly due to the distributed nature of those control-loops, which requires global knowledge of the event-action interactions. Furthermore, if control-loop entanglement is to be considered, the choreography aspect gets even more dominant, when it comes to specifying the precise, and timely, interactions of the participating DEs. A particularly effective approach for analyzing and verifying these control scenarios is the application of hierarchical CPNs, which help formally specify and validate stability and scalability properties of GANA compliant autonomic behavior. One of the most stringent entanglement problem is dead-lock detection within network level control-loops, which is formally analyzable with PNs tools such as *CPN Tools* [24,25,26,27]. The enabling meta-model snippet is shown in Figure 23.

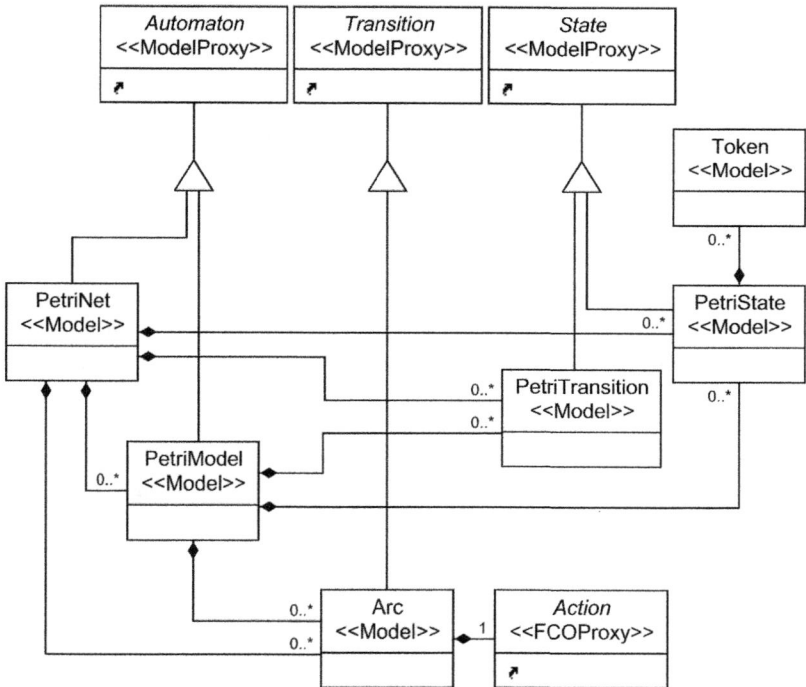

Fig. 23. Meta-Model of Petri-Nets

The GANA meta-model is a complex modular meta-model, which takes advantage of all the advanced features of MIC, though staying compatible with MOF, conceived with the intent to support detailed use case control scenario models according to the guidelines and explicit architectural constraints of the GANA reference architecture. By this meta-model facility the GANA reference model becomes formal and analyzable against stability, performance and scalability properties. The meta-model also incorporates a novel *model-patent* pattern that enables the creation of GANA models by providing a maintainable and highly modularized modeling infrastructure for dealing effectively with repeatedly reused and incrementally and iteratively enhanced modeling elements that formally specify the major components of highly distributed autonomic protocol-driven control systems.

6   Behaviour Modeling and Design of Autonomic Entities

We assume that every protocol and ME of an autonomic network is managed by an advanced autonomic controller. The first step required before designing the behavior of such an advanced autonomic controller is the model of the protocols' and MEs' behavior they are assigned to manage. Describing a certain behavior of a protocol or ME through *classical first-principles* modeling techniques is difficult and complex, and requires a detailed knowledge of the relationships between the inputs and outputs. Manually constructing the higher order difference equations for Multiple-Input-Multiple-Output (MIMO) discrete time systems (such as protocols and MEs) with several inputs and outputs is not only tedious but also time consuming. However, interpreting the behavior of protocols and MEs using de facto FSM techniques, while easy, would be unproductive from the stability analysis [9] point of view, as this representation presents only a limited set of opportunities. For instance it would be nearly impossible to measure the boundness of the output with respect to the boundness of the input with FSM representation. Further, other properties such as accuracy (steady-state error), settling time, and peak overshoot cannot be conveniently analyzed with FSMs. On the other hand, FSMs are simple, easy to design and analyze and most importantly flexible. Thus there is a need for a hybrid approach.

Due to these contrasting advantages of FSMs and *first-principles* models, we employ a hybrid approach to model and design the behavior protocols, MEs and their autonomic controllers respectively. *System Identification* techniques defined in the field of modern control theory provide concise means of constructing a system's behavior model as *difference equations*. Thus, for a computer network based on hierarchical autonomic network architecture, such as GANA described in 2.1, we enlist this hybrid approach. The Level-2 DEs and the Level-1 protocols and MEs are designed and modeled using techniques from modern control theory (feedback control or state-space feedback control) for system modeling and controller design. The behavior of higher levels, Level-3 and Level-4 DEs are designed as FSMs or PNs. This is illustrated in Figure 24. Before we proceed with the details of this method, we assume that a structural model of the targeted autonomic network is available as input to this method.

6.1   Modeling the Behavior of Protocols and Managed Entities

For the purpose of modeling the behavior of protocols and MEs, the modeling techniques from the modern control theory domain are exploited. The methods

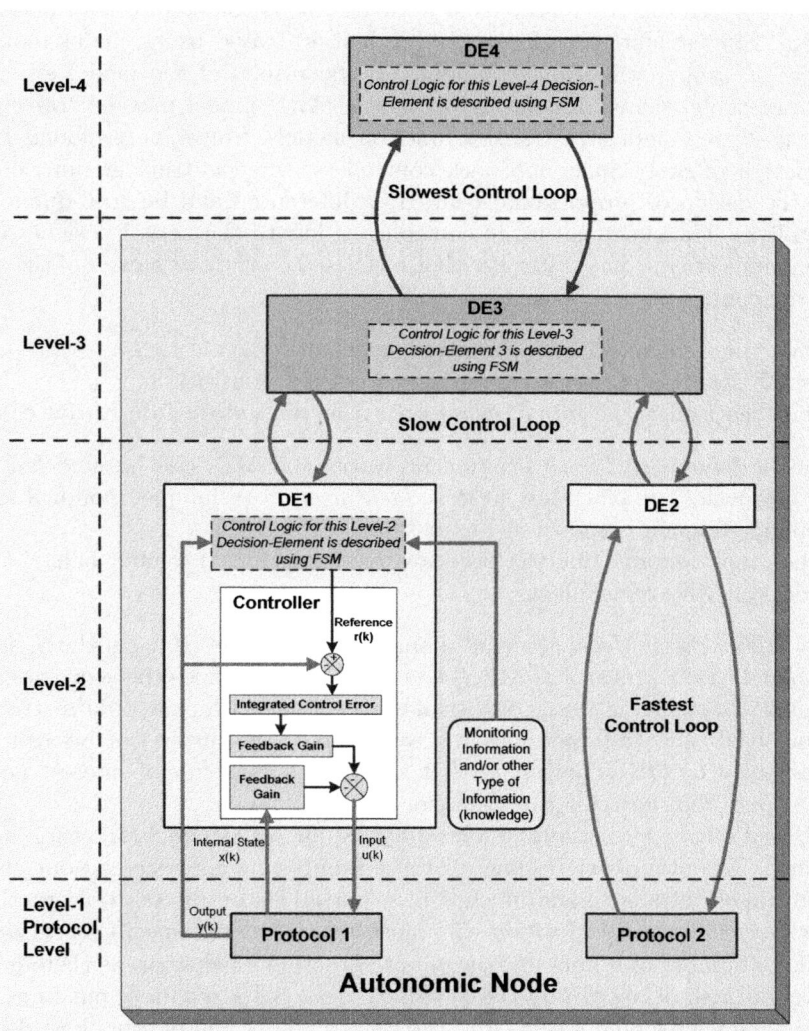

Fig. 24. Controller Design using both FSMs and Feedback controller approach

described in [8] are applicable to domains of robotics and industrial automation. We adapt these methods [8,7] to the domain of autonomics and networking.

The behavior of most protocols and computing systems (MEs) can be modeled as some form of *AutoRegressive eXogenous* (ARX) model [7, pp 38–40]. An ARX model is a linear difference equation that equates the input $u(k)$ and output $y(k)$ as follows $y(k+1) = ay(k) + bu(k)$ in its simplest form. As *first-principles* models are harder to realize manually, statistical and black-box *System Identification* techniques are applied. In black-box *System Identification*, experiments are be designed in order to gain statistical information of the system under observation. This information is used to infer (estimate) parameters of the the abstract ARX model representing the relationship between the various inputs and outputs of system. The estimated model would be then evaluated using one or more statistical accuracy evaluation techniques. If the results of the model evaluation are acceptable, the estimated and evaluated ARX model may be transformed to State-Space models or transfer function models. State-Space models enable the design of State-Space feedback controllers, whereas transfer functions enable the design of proportional (-integral, differential and integral-differential) controllers. The exact nature of controller is left to the actual designer of the autonomic entities. The controller choice is based on the complexity of the model and the control objectives such as:

- fast transient response (proportional control),
- small steady-state error (proportional-integral control), and
- fast reaction to the *rate of change* of error (proportional-derivative control).

Thus the behavior of both existing protocols and MEs can be modeled using this approach. The black-box *System Identification* technique, modified for the autonomic domain is shown in Figure 25.

The steps required for the accurate modeling of a targeted behavior of a protocol or ME are as follows:

1. The *Structural Model* of an autonomic network is parsed to get the *Structural Model* of the protocol or ME. The parsed model defines the structure of the protocol or ME in terms of the interfaces, relationships and data structures involved, and thus provides the inputs and the outputs of interest for a protocol or ME. The inputs that influence a behavior of interest and the outputs that describe this behavior are then chosen.
2. Based on the chosen inputs and outputs, an ARX model structure is specified. To obtain best results, all those inputs that influence an output or a group of outputs should be chosen. A partial set of inputs would result in a *underfitting* model. However care must be taken not to specify an excessively large number of inputs and outputs to describe a behavior, as there is a potential risk of *overfitting* the model. A trade off is required, but in general, the modeler should ensure that the chosen inputs and outputs best describe the targeted behavior. This information is usually available in literature or in some survey.
3. Experiments are then designed to collect statistical information regarding the inputs and outputs. The collected data is used to estimate the parameters of the ARX model. The experiments should be designed such that:
   - the entire range of valid values for an input are used, and
   - the various values of the inputs are chosen in such a way that they provide a good coverage of the *inputs space*, and finally
   - the chosen values are able to excite the entire range of behavior dynamics in the protocol or ME.
4. Parameter estimation involves the use of statistical techniques to estimate the parameters of the model. A frequently used technique is the *least squares*

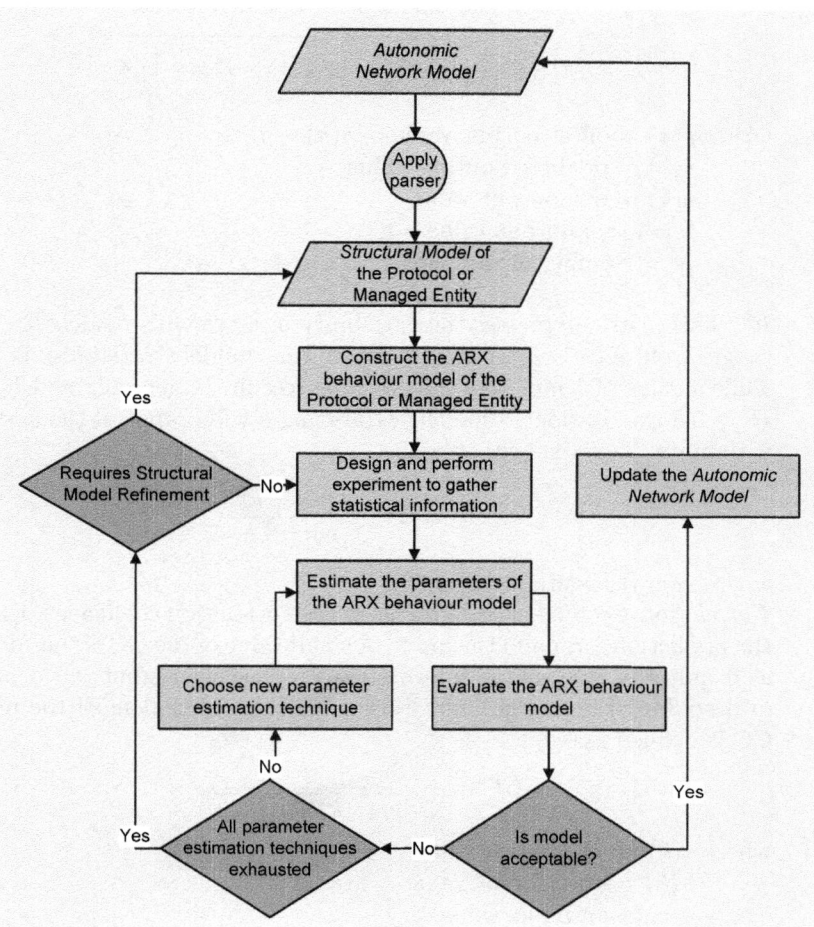

Fig. 25. Modified black-box *System Identification* from [7, pg. 44] for modeling the behavior of Protocols and Managed Entities of an Autonomic Network

regression (LSR) technique. However, other statistical estimation techniques should be studied and used, in cases when the model whose parameters were estimated using LSR technique provide inconclusive or poor evaluation results. A detailed treatment of the various estimation techniques is beyond the scope of this paper. [28] provides a survey of various techniques for *System Identification* through parameter estimation.

5. Evaluation of the estimated behavior model reveals the depth and validity of the experiment. [7, pp. 53-54] describes three metrics that can be used for assessing the verity and validity of the estimated model. As defined in [7] they are:

   – *Root-Mean-Square Error (RMSE)*: This metric estimates the standard deviation of the residuals and thus provides insights into the accuracy of the estimations used in the model. RMSE is defined as :

$$RMSE = \sqrt{\frac{1}{N} \sum [y(k+1) - \hat{y}(k+1)]^2} \qquad (3)$$

where $y(k)$ = offset output value = $\tilde{y}(k) - \bar{y}$
$\hat{y}(k)$ = predicted output value
$\tilde{y}(k)$ = raw output value
$\bar{y}$ = mean output value
N = Number of samples

- R^2: This metric expresses the variability in a system's model. R^2 value ranges from 0 to 1. A value of 0 means that model's variability is poor, while a value of 1 indicates that its a perfect fit. In general, models with $R^2 \geq 0.8$ are considered models expressing a wide range of the system's variability [7]. R^2 is defined as:

$$R^2 = 1 - \frac{var(y - \hat{y})}{var(y)} \qquad (4)$$

where $var(y)$ = variance of $y(k)$
- *Correlation Coefficient (CC)*: This is the correlation coefficient between the prediction error and the input. A small value of the CC is thus desired as it indicates that all the information supplied by a input was captured in describing the model, indicating no further refinement of the model. CC is defined as:

$$CC = \frac{\sum_{k=1}^{N} e(k)u(k)}{\sqrt{var(e(k))var(u(k))}} \qquad (5)$$

where $e(k)$ = offset error value = $y(k) - \hat{y}(k)$
$u(k)$ = offset input value = $\tilde{u}(k) - \bar{u}$
$\bar{u}$ = mean input value

In addition to these metrics, a scatter plot projecting *measured values* versus *predicted values* may provide an added insight to the accuracy of the parameters and the models. This is because, for instance, even though a large value of R^2 is supposed to express the system's variability in the model, such large values can be also obtained if the measured values are clumped around the fringes of an operating region of the system.

6. Once the model is considered to be *fit*, the behavior model of the protocol or ME is integrated with its structural model, and the autonomic network model updated. Else, the above steps are iterated once again till a model of considerable fitness is obtained.

In some special cases when experiments cannot be designed or when the accuracy of the model is poor or unreliable, a FSM (or one of its many variations) representing the behavior of the protocol may be used. While the methodology is flexible to accommodate this, certain aspects of the formalized model analysis would be unavailable.

6.2 Designing the Behavior of Level-2 Decision-Elements

The control-loops operating over protocols and MEs are the fastest control-loops in autonomic networks with hierarchical control-loops. Thus, they operate with concrete reference objectives and policies over the protocols and MEs they intend to control. In such control-loops a direct manipulation of the inputs of the protocol or ME is carried out based on the feedback of the observed output. This is illustrated in Figure 24.

As the behavioral modeling of protocols and MEs, described in section 6.1, are carried out using the modeling techniques from modern control theory, it is imperative that the same techniques are used for the design of the Level-2 advanced autonomic controllers or DEs. However, the control-loops operating above Level-2 DEs, i.e., the *Level-3 - Level-2* control-loop are slow control-loops with a mixture of both abstract and concrete policies and objectives. The design of such control-loops and their higher control-loops could follow the conventional protocol design techniques, by expressing the behavior models through FSMs. This means that a behavior model of a Level-2 DE's consists of two different design formalisms. However, as DEs are not just limited to the role and functionality of a controller, their behavior model described as a FSM, could be stacked on top of the controller designed as a *State-Space Feedback Controller* or a *Proportional Controller* (or its various other forms). This behavioral model stacking is visible in Figure 24. This means that the FSM stacked on top of the controller would essentially drive the controller by providing it with concrete reference inputs. The exact values of these reference inputs provided to the controller depend on the current state of the FSM, the output from the protocol's or ME's behavior and other information such as monitoring information.

Figure 26 illustrates the steps required for an accurate behavior design of a Level-2 DE. They are as follows:

1. The model of the protocol or ME of interest is chosen from the *Autonomic Network Model*. The parsed model provides both the structural and behavioral model of the protocol or ME.

2. An appropriate controller design such as *Proportional Controller* or *PID Controller* or *State-Space Feedback Controller* is chosen based on type of behavioral model (state-space model or transfer function), and on the type of control objectives desired for a particular behavior of the protocol or ME. The properties of interest during the controller design for autonomic networks are: **S**tability, **A**ccuracy, **S**ettling Time and **M**aximum Peak **O**vershoot. These **SASO** properties must be satisfied for the proper design of a controller and effective operation of the protocol or ME under control.

3. The parsed *Structural Model* of the Level-2 DE is then used to describe the behavior of the *Level-3 - Level-2* control-loop. The FSM designed for this purpose is stacked on top of the controller designed in the previous step. The FSM controls the reference values (objectives) of its underlying controller.

4. The FSM is then evaluated for its *fitness* and if acceptable, both the controller and the FSM are integrated to the *Autonomic Network Model* as behavior models of the Level-2 DE.

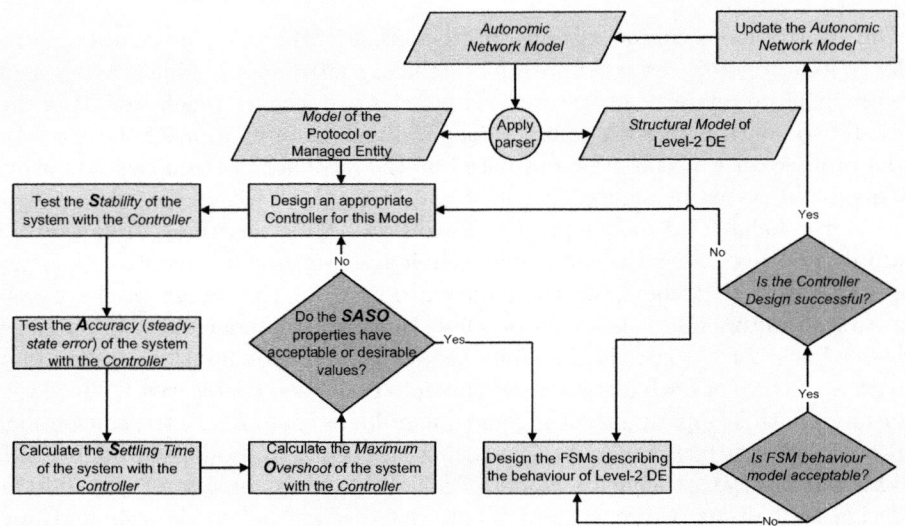

Fig. 26. Steps to design the behavior of Level-2 Decision-Elements of an Autonomic Network with Hierarchical Control-Loops

In some special cases, the FSM-Controller behavior model stack can be replaced by a single FSM, if and only if the behavior model of the protocol or ME is described as a FSM. Thus the *Behavior Modeling and Design* method is flexible. However, evaluation of certain properties (like *SASO*) would be almost impossible under such circumstances.

6.3   Designing the Behavior Level-3 and Level-4 Decision-Elements

The Level-3 and Level-4 DEs operate in a distributed fashion with abstract goals and policies. The *Level-4 - Level-3* control-loop is the slowest control loop in the autonomic architecture. These DEs abstract the behavior of networking functions, and are as such built for optimizing these abstracted functionalities. Thus their policies and objectives are defined as utility functions [29], with the *optimal point* set by the operator/administrator. Designing the behavior of such Level-3 and Level-4 DEs can be carried out through Communicating Finite State Machines (CFSMs), PNs or Message Sequence Charts (MSCs). Each of these design formalisms come with their advantages and drawbacks and their own tooling infrastructure. The exact choice of behavior formalism is dependent on the designer of these DEs.

The *Behavior Modeling and Design* method provides a possibility for the realization of a decentralized and distributed behavior modeling and design approach. Individual modelers and designers of protocols or MEs and DEs, can work independently of each other and the behavior formalism without worrying about its impact on the upper DEs or underlying protocols or MEs. The other benefit is that it masks the drawbacks of each individual behavior specification formalism by complementing it with advantages of the other. For instance, design techniques from control theory ensures that the protocol and its immediate controller satisfy the *SASO* properties. However, the perceived drawback of this approach is the complexity and time required to design and perform these data aggregating experiments. Applying the same design approach for the upper-level DEs would not only be extensively tedious, but also complex and time consuming. This drawback of control theory based design techniques can be masked and replaced by the advantages provided by FSMs, PNs and MSCs. Describing the behavior of the upper-level DEs through FSMs, PNs or MSCs are simpler and faster, and pave a way to formally verify, validate and simulate the behavior of these autonomic components. Its worth emphasizing once again that the glue here is the *meta-model* and the *Methods Integration Framework*, that allows the designer to describe and exchange the structure and behaviour models of several autonomic entities.

In this way, the methodology and its formal methods blend the best of many dissimilar designing worlds for modeling and designing conventional protocols, managed entities (MEs) and components (DEs) of an autonomic network.

7   Case Study

In this section, we document the details and results of a case study conducted to validate and showcase the features of our methodology. The case study chosen for this purpose is taken from the scenario documented in [30]. The scenario involves the use of risk information to control the routing behavior of the OSPF protocol, by manipulating the link-costs on the interfaces of a router, with respect to the variation of the risk levels in the router/node. The risk-level of a router is defined as a function of the operating temperature of a router. Thus, as the operating temperature of a router goes above a set threshold value, the router is considered risky, and the link-cost on the interfaces of the router are increased to discourage traffic through this router. On the other hand, when the operating temperature falls below the threshold value, the link-costs on the router interfaces are lowered to allow more traffic to flow through the router.

As the purpose of this case study is to showcase the processes of the individual formal methods, we have modified the scenario, such that we now use the risk information to control the values of *HelloTimers* of OSPF, and thus control the OSPF protocol traffic (not the data traffic). The relationship between the values of the *HelloTimers* (and *RouterDeadInterval*) and OSPF routing traffic is intuitively straight forward, and can be seen as a inverse relationship. This means that, a higher value of *HelloTimers* (and *RouterDeadInterval*) results in lower control traffic between the routers, and vice-versa. We showcase, how this intuitive relationship (RFC2740 [31]), is the same when we perform the black-box *System Identification* experiments to determine this relationship. The purpose of this case study is for pedagogical analysis only. The experiments were conducted in the OPNET [32] simulation environment, and thus the experimental results obtained are tied to the OSPF implementation of the simulator.

7.1   Structural Modeling of Decision-Elements

Section 5 describe the methods for meta-modeling and structural modeling of the autonomic components. Figure 27 depicts the meta-model of the GANA Decision-Elements (DEs). The GME [22] tooling infrastructure is used for formalizing the GANA meta-model. As it may be seen, the meta-model formalizes the specifications of the DE, namely its interfaces, its behavior formalisms, its autonomic mechanisms, its data-types and so on. The fundamental feature of the GANA meta-model is the control-loop. The control-loop specifies which elements are involved in the control-loop and their relationships to the other components of the autonomic network. Once the meta-model is well defined, it is interpreted and registered as a new paradigm. The new paradigm allows models to be described in the domain of the autonomic architecture, such as GANA. Figure 28 shows such a structural model, namely the *GANACaseStudy*.

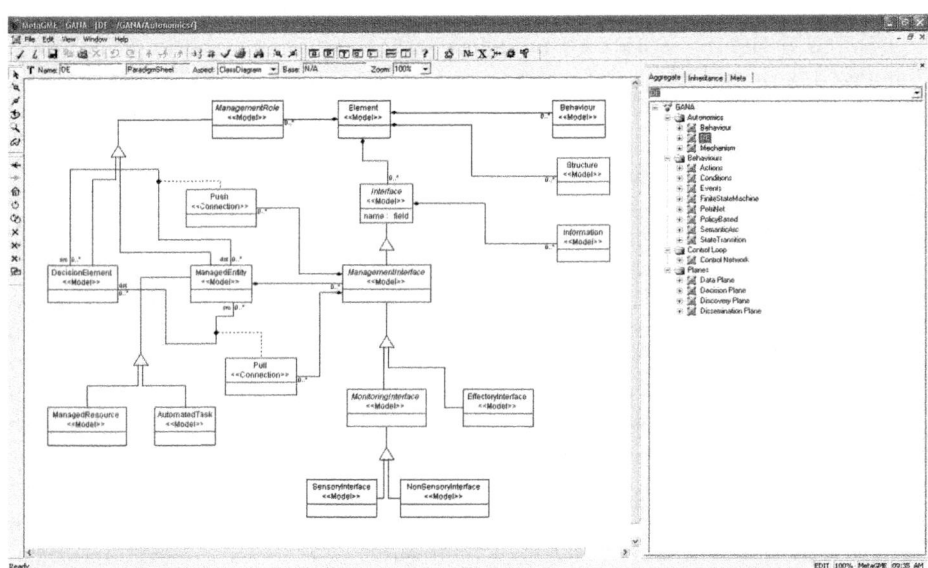

Fig. 27. GANA Meta-Model - Meta-Model of GANA Decision-Element

As it may be seen, a control-loop named *RiskAwareRoutingControl*[1] is defined, and consists of two participating elements *GANANode*[2], and *Network-Level-Element*[2]. The *GANANode* in turn consists of the structural models of the NODE_MAIN_DE[2], FUNC_LEVEL_RM_DE[2] and the OSPFv3 protocol[3]. The formalism used for modelling the individual behavior of DEs and OSPF is documented in Table 1.

---

[1] Instance of *Control-Loop* meta-model described in Section 5.1.1.

[2] Instance of *Decision Plane* meta-model described in Section 5.1.2.

[3] Instance of *Protocol* meta-model described in Section 5.1.3.

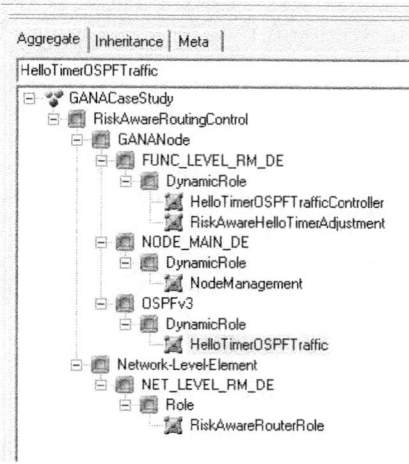

Fig. 28. Model of Case Study showing the Hierarchy of the DEs

Table 1. Behavior Formalisms used for describing the behavior of GANA Decision-Elements and Protocols/Managed-Entities

Autonomic Entity	Behavior Formalism	Model Name
NET_LEVEL_RM_DE	Moore Automaton	RiskAwareRouterRole
NODE_MAIN_DE	Moore Automaton	NodeManagement
FUNC_LEVEL_RM_DE	Moore Automaton Proportional Controller	RiskAwareHelloTimerAdjustment HelloTimerOSPFTrafficController
OSPFv3 Protocol	Difference Equations	HelloTimerOSPFTraffic

The *Structural Model* defined here is used by the *Behavior Modeling and Design* method to model and design the behavior of OSPF and DEs respectively.

7.2 Behavioural Modeling of OSPF

The *Structural Model - GANACaseStudy*, defined in section 7.2 contains the structure of the OSPF protocol, i.e., the interface parameters. These parameters can be categorized as inputs and outputs and are chosen by the designer to describe the behavior of the protocol. As part of our case study, we chose the *HelloTimers* parameter as the input and *OSPFTraffic* as the output. In order to model the effect of *HelloTimers* on *OSPFTraffic* (OSPF behavior), we follow the method described in Section 6.1.

We consider a *first-order model*, expressed as an ARX model below:

$$y(k + 1) = a \cdot y(k) + b \cdot u(k) \tag{6}$$

where $y(k)$ = Variation of OSPFTraffic (Kb/s) with time(k)

$\quad\quad u(k)$ = Variation of HelloTimers value with time(k)

$\quad\quad a, b$ = Scalars

We are interested in the values of a and b. For this purpose, we designed an experiment measuring the *OSPFTraffic* for various values of *HelloTimers*. The results are document in Table 2.

Table 2. Results from the Experiment

Sample k	HelloTimers (s) $\tilde{u}(k)$	OSPFTraffic (Kb/s) $\tilde{y}(k)$
1	10	73
2	12	62
3	14	53
4	16	48
5	18	42
6	20	38

In order to estimate the parameters a and b, we use the LSR technique. The mean values for *HelloTimers* are calculated over samples $[1, 5]$, as the last value is not used in the parameter estimation calculation. The mean values for *OSPFTraffic* is calculated over the samples $[2, 6]$. The values are:

$$\bar{u} = 14 \quad and \quad \bar{y} = 48.6 \tag{7}$$

The offset values $u(k) = \tilde{u}(k) - \bar{u}$ and $y(k) = \tilde{y}(k) - \bar{y}$ are recorded in Table 3.

Table 3. Calculated Offset Values

k	$\tilde{u}(k)$	$\tilde{y}(k)$	$u(k)$	$y(k)$
1	10	73	-4	24.4
2	12	62	-2	13.4
3	14	53	0	4.4
4	16	48	2	-0.6
5	18	42	4	-6.6
6	20	38	6	-10.6

a and b are estimated with LSR technique [7] as follows:

$$a = \frac{S_3 S_4 - S_2 S_5}{S_1 S_3 - S_2^2} \quad\quad b = \frac{S_1 S_5 - S_2 S_4}{S_1 S_3 - S_2^2} \tag{8}$$

where

$$S_1 = \sum_{k=1}^{N} y^2(k) \quad S_2 = \sum_{k=1}^{N} u(k)y(k) \quad S_3 = \sum_{k=1}^{N} u^2(k)$$

$$S_4 = \sum_{k=1}^{N} y(k)y(k+1) \quad S_5 = \sum_{k=1}^{N} u(k)y(k+1)$$

(9)

Tool such as MATLAB [33] or Octave [34] can be used to calculate the parameters defined in equation 9. The calculated values for these parameters are listed below.

$$S_1 = 838.2 \quad S_2 = -152 \quad S_3 = 40 \quad S_4 = 457.2 \quad S_5 = 118 \tag{10}$$

Applying these values to equation 8, the values obtained for a and b are:

$$a = 0.033768227 \approx 0.03 \quad and \quad b = -2.821680737 \approx -2.82 \tag{11}$$

Substituting these values to equation 6, the model protocol's behavior of interest is:

$$y(k+1) = 0.03y(k) - 2.82u(k) \tag{12}$$

Further, with values of a and b, we calculate the predicted output value $\hat{y}(k+1)$, given as:

$$\hat{y}(k+1) = a \cdot y(k) + b \cdot u(k) \tag{13}$$

The model is then evaluated according to evaluation requirements given in equations 3, 4 and 5. The results are:

$$RMSE = 1.360 \qquad R^2 = 0.981 \qquad CC = 4.06 \times 10^{-15} \tag{14}$$

All these values indicate that the model has few errors, is accurate, and require no further refinement. A scatter plot projecting measured values versus predicted values is shown in Figure 29. As it may be seen, the measured values are are close to the values predicted by the model, indicating the accuracy of the model.

7.2.1   Model Analysis

In order to analyze the generated model, we use the Z-transformation of model. The Z-transform of equation 12 is:

$$G(z) = \frac{-2.82}{z - 0.03} \tag{15}$$

All the poles of the system are inside the unit circle of the z-plane. This indicates that the system is BIBO stable. As the absolute value of the pole is far away from the unit cicle, it indicates that the *Settling Time* of the system is low. As we used increments of $2s$ to sample *HelloTimers*, a step input of amplitude 2 (Z-transform $\frac{2z}{z-1}$) would result cause the system to reach its steady-state in $\approx 2.62s$, which is as expected. Further, as $|a| \approx 0$, it indicates that there is very little system dynamics. On the other hand, $b < 0$, indicating that the input

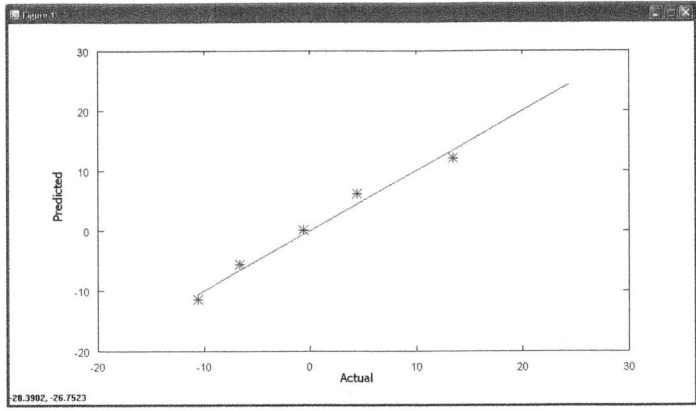

Fig. 29. Scatter Plot - Predicted Values vs. Actual Values

and the output are inversely related to each other. Thus the black-box *System Identification* technique generates a model that is close to the intuitive model predicted before starting this case study, showcasing the power of such techniques to determine a system model when they cannot be manually constructed. The impulse-response and the step-response of the system is shown in Figures 30 and 31 respectively.

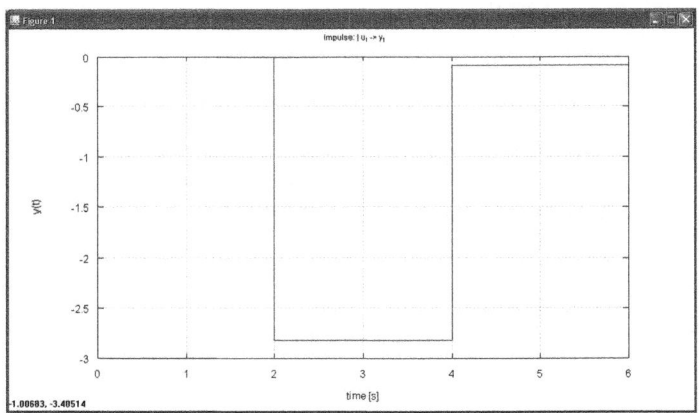

Fig. 30. Impulse Response of the Behavior Model

The impulse response shows how the OSPFTraffic would behave if the *HelloTimers* were manipulated as impulses. Thus, if the *HelloTimers* were varied as impluses, the traffic would momentarily drop down, before going back to its steady-state value. Thus, the variation of *HelloTimers* as impluses to the system

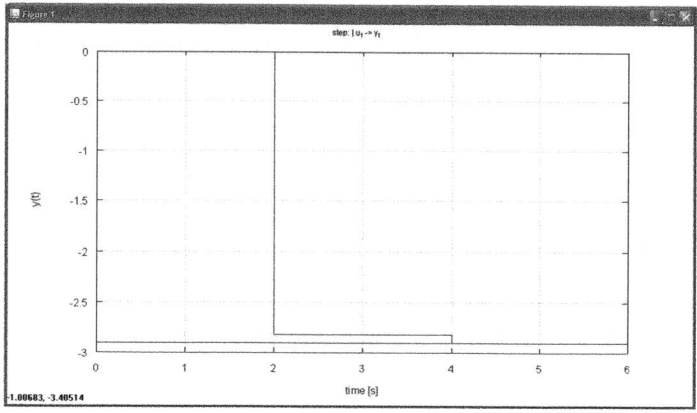

Fig. 31. Step Response of the Behavior Model

should be discouraged to avoid steady-state oscillations, which in our case is the OSPF's control traffic. On the other hand, varying the *HelloTimers* as a step input would gradually increase or decrease the OSPF control traffic. It should be noted that in both the cases, the output reaches its steady-state value after a certain delay. This should be taken into account, while designing its controllers and other higher level DEs.

7.3 Designing Level-2 Controller

A Level-2 DE's controller design depends on the control-objectives of a system's behavior, the complexity of the behavior model, and the requirements of the autonomic architecture. Based on the model obtained in section 7.2, we choose the *Proportional Controller*. The block diagram of a feedback control system with *proportional control* is shown in Figure 32.

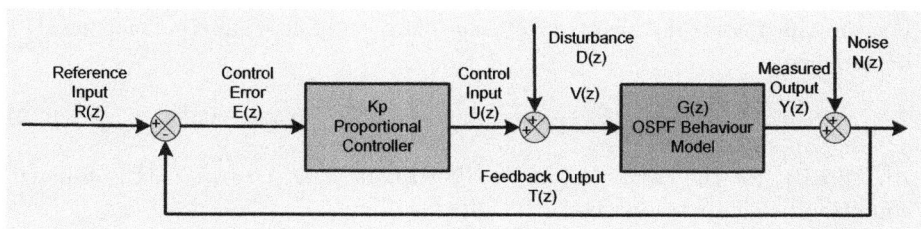

Fig. 32. Feedback Control System with Proportional Control

As the name suggests, a *Proportional Controller* controls the input to a system proportional to the control error value. The design of a *Proportional Controller* and its various variants is based on the pole placement design technique, where the poles of closed-loop system are chosen by desire to meet the control objectives. The values of the *SASO* properties are the control objectives. Thus for our case study, we set the following pedagogical control objectives.

1. The system needs to be *Stable*.
2. *Accuracy* or Steady-State Error $e_{ss} < 0.1$
3. *Settling Time* $k_s < 2s$
4. *Maximum Peak Overshoot* $M_p < 0.15$

The closed-loop transfer function is given as

$$F_R(z) = \frac{Y(z)}{R(z)} = \frac{G(z)K_p}{1 + K_pG(z)} \tag{16}$$

Thus for our system it is:

$$F_R(z) = \frac{-2.82K_p}{z - 0.03 - 2.82K_p} \tag{17}$$

The poles of the closed-loop transfer function given in equation 17 are: $p_1 = 0.03 + 2.82K_p$. For a system to be stable, the poles of the system must be inside the unit circle of the complex plane. Thus $|P_1| = |0.03 + 2.82K_p| < 1$. Solving this gives,

$$- 0.365 < K_p < 0.343 \tag{18}$$

Moving on to the second control objective, the steady-state error is desired to be < 0.1. For this objective to be met, we need to satisfy the following condition:

$$e_{ss} = r_{ss}[1 - F_R(1)] \tag{19}$$

where $r_{ss} = 1$ since we are using a unit-step reference value ($F_R(1)$). Solving this, we get

$$e_{ss} < 0.1 \quad \Rightarrow 1 - \frac{-2.82K_p}{1-0.03-2.82K_p} < 0.1 \quad \Rightarrow K_p < -0.309 \tag{20}$$

For the third control objective *Settling Time*, the following condition needs to be satisfied:

$$k_s \approx \frac{-4}{log|0.03+2.82K_p|} < 2 \quad \Rightarrow 0.037 < K_p < -0.058 \tag{21}$$

And finally, for the control objective *Maximum Peak Overshoot*, the following condition needs to be satisfied:

$$M_p < 0.15 \quad \Rightarrow M_p = |0.03 + 2.82K_p| < 0.15 \quad \Rightarrow -0.06 < K_p < 0.042 \tag{22}$$

To summarize, the values of K_p for the various control objectives are as follows:

1. *Stability*: $-0.365 < K_p < 0.343$
2. *Accuracy*: $K_p < -0.309$
3. *Settling Time*: $0.037 < K_p < -0.058$
4. *Maximum Peak Overshoot*: $-0.06 < K_p < 0.042$

The values of K_p obtained for the various control objectives is plotted in Figure 33. As it may can be seen, not all control objectives can be satisfied by a particular value of K_p. Choosing K_p values between $[-0.365, -0.309]$ would satisfy, the *Stability*, *Accuracy* and *Settling Time* objectives at the expense of the *Maximum Peak Overshoot*. Choosing K_p values between $[-0.060, -0.058]$ or $[-0.037, -0.042]$ would satisfy the all objectives with the exception of *Accuracy*. Thus *Accuracy* and *Maximum Peak Overshoot* provide conflicting requirements on the choice of K_p. Its up to the individual designer or the autonomic architecture's control strategy requirements, to decide the trade off between these two control objective requirements. A relaxation of either objectives' requirements is another possible solution.

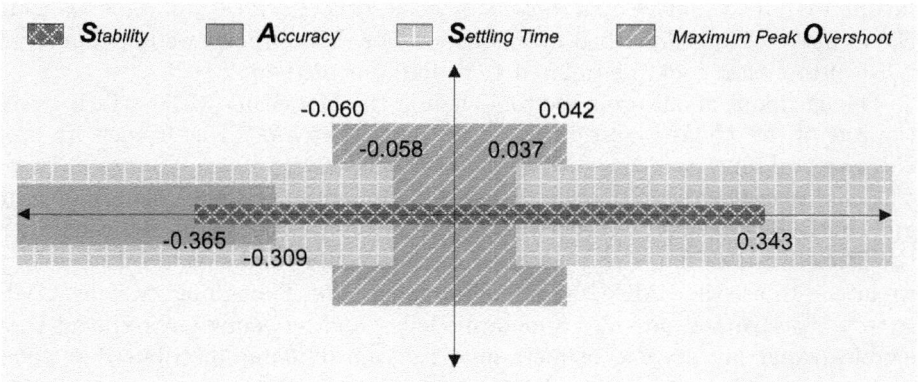

Fig. 33. Ranges of K_p values obtained for SASO control objectives

7.4   Designing Level-2, Level-3 and Level-4 Decision-Elements

The DEs of the scenario are modeled as FSM. In accordance to GANA's design requirements, the state machines of the Level-2 and Level-4 DEs are mirror state machines. However, they operate on different information sets (and thus different state-transitions conditions). The Level-3 DE, i.e., the NODE_MAIN_DE is essentially a node management DE, bootstrapping, orchestrating and (re)configuring the node's protocols and Level-2 DEs. The state-machines for the Level-2 and Level-4 DEs are depicted in Figures 34a and 34b respectively.

8   Future Research Work

In the future, we intend to extend the current version of the meta-model with additional specifications from the GANA reference architecture. The methods currently presented were individually validated. The absence of a methods integration framework meant that the models that were developed by the individual methods could not be formally integrated into a single model asset. Further, the

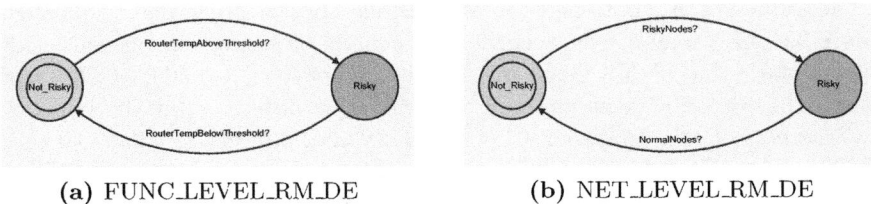

(a) FUNC_LEVEL_RM_DE   (b) NET_LEVEL_RM_DE

Fig. 34. Behaviour Models of Level-2 and Level-4 Decision-Elements described as Finite State Machines

purpose of this research work was the highlight the strengths of the individual methods and their benefits to the domain of autonomic networking. Thus, in the future, we intend to develop a tool-chain, consisting of a variety of tools realizing the formal methods described in this paper. For this purpose, we have surveyed a list of tool that could be tailored to fit into our methodology.

The backbone of our proposed tool-chain is the ModelBus [35,36] which serves the role of the *Methods Integration Framework*. The ModelBus framework provides several functionalities such as *model fragmentation, model change notification* and *workflow orchestration*. These functionalities are used for exchanging and controlling model fragments among the various tools that are integrated with the ModelBus [36]. For the purpose of meta-modeling and structural modeling, we intend to use the GME [22] tooling infrastructure. Though lightweight, GME is very effective and provides a meta-modeling and modeling environment that enables rapid prototyping of meta-model driven dynamic distributed systems such as [37]. The behavior models would be specified using tools such as UP-PAAL [38] for FSMs, CPN-Tools [39] for PNs and MATLAB [33] or OCTAVE [34] for modern control theory based numerical computations. In addition to these, we intend to use Microsoft Visio [40] to express our scenarios as MSCs, which could in turn be translated to FSM adopting the methodology reported in [37], using the M2Code Tool [41]. The envisioned tool-chain is depicted in Figure 35.

In this case study, we have currently modeled a Single-Input-Single-Output (SISO) behavior. In the future, we intend to model a multiple-input behavior. This means that the current behavior would be extended and the influence of other OSPF parameters, such as *LSRefreshTime, MinLSInterval* and *MinLSArrival* on OSPF control traffic will be investigated. Further, we intend to use the OMNeT++ [42] network simulation environment, as this allows the porting of software based routers such as Quagga [43] into their simulation space. Thus the models generated would be models of the real OSPF implementations available in the industry. Further, the current black-box System Identification techniques used for our purposes are applicable only to linear systems. The authors of [44] have provided a genetic engineering approach to system identification of non-linear systems. We intend to investigate further in this direction. Finally, the extended model is intended to be designed, refined and verified using the envisioned tool-chain.

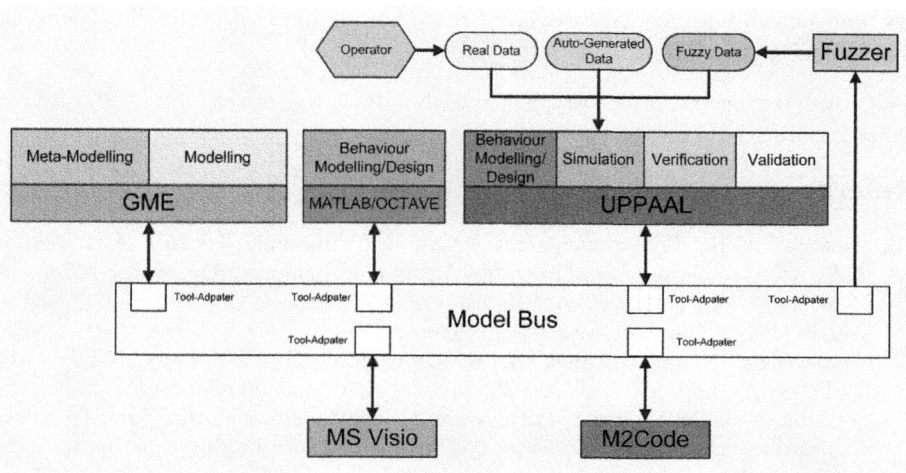

Fig. 35. Envisioned Tool-Chain for realizing the Methodology

9 Conclusion

In this paper, we provided formal methods to design, refine and verify the behaviour of current network protocols and new autonomic components built of top of them as autonomic managers. The motivation for providing these formal methods stems from the fact that the domain of autonomics and other nature-inspired networking solutions come with their own set of challenges and requirements, that need a tailored solution rather than a generic one. In our approach, we split the modeling of autonomic components into structural modeling and behavioral modeling. This follows from our modeling and design philosophy - *"skeletons first, organs next"*. Nevertheless, understanding the pragmatic problems of specification and modeling, we relax this Utopian approach by allowing the refinement of the structural models after the behavior models have been specified.

In order to showcase the practical benefits of our methodology, we conducted a case study instantiating the components of a GANA based autonomic network. The results of the case study provide a promising reinforcement for the requirement of such formal methods to design and verify autonomic components. The insights provided by the case study indicate that the domain of autonomics cannot be catered by conventional system's engineering approaches. Nevertheless, the work done here is by no means sufficient. We have identified areas where further work needs to done, for instance in developing a tool-chain for realizing the methodology.

In conclusion, designing networks based autonomic or nature-inspired networking architectures is not trivial. It requires the use of modeling, designing, verifying, validating and simulation methods/techniques from a wide range of domains like protocol design techniques, system engineering, modern control theory, and possibly more. An associated tool-chain is imperative to the success of this hybrid methodology.

Acknowledgment. This work is partially supported by EC FP7 EFIPSANS project (INFSO-ICT-215549) [45].

References

1. Chaparadza, R.: Requirements for a Generic Autonomic Network Architecture (GANA), suitable for Standardizable Autonomic Behavior Specifications for Diverse Networking Environments. International Engineering Consortium (IEC), Annual Review of Communications, 61 (2008)
2. Chaparadza, R., Papavassiliou, S., Kastrinogiannis, T., Vigoureux, M., Dotaro, E., Davy, A., Quinn, K., Wódczak, M., Toth, A., Liakopoulos, A., Wilson, M.: Creating a viable Evolution Path towards Self-Managing Future Internet via a Standardizable Reference Model for Autonomic Network Engineering. In: Towards the Future Internet - A European Research Perspective, pp. 313–324. IOS Press (2009)
3. Rétvári, G., Németh, F., Chaparadza, R., Szabó, R.: OSPF for Implementing Self-adaptive Routing in Autonomic Networks: A Case Study. In: Strassner, J.C., Ghamri-Doudane, Y.M. (eds.) MACE 2009. LNCS, vol. 5844, pp. 72–85. Springer, Heidelberg (2009)
4. Bell, J.: Understand the Autonomic Manager Concept (October 2004),
   http://www.ibm.com/developerworks/autonomic/library/
   ac-amconcept/index.html
5. Greenberg, A., Hjalmtysson, G., Maltz, D.A., Myers, A., Rexford, J., Xie, G., Yan, H., Zhan, J., Zhang, H.: A Clean Slate 4D Approach to Network Control and Management. SIGCOMM Comput. Commun. Rev. 35(5), 41–54 (2005)
6. Ballani, H., Francis, P.: CONMan: A Step Towards Network Manageability. In: SIGCOMM 2007: Proceedings of the 2007 Conference on Applications, Technologies, Architectures, and Protocols for Computer Communications, pp. 205–216. ACM, New York (2007)
7. Hellerstein, J.L., Diao, Y., Parekh, S., Tilbury, D.M.: Feedback Control of Computing Systems. John Wiley & Sons Inc., Hoboken (2004)
8. Ogata, K.: Modern Control Engineering. Prentice Hall PTR, Upper Saddle River (2001)
9. Prakash, A., Chaparadza, R., Theisz, Z.: Requirements of a Model-Driven Methodology and Tool-Chain for the Design and Verification of Hierarchical Controllers of an Autonomic Network. In: Third International Conference on Communication Theory, Reliability, and Quality of Service (2010)
10. Tcholtchev, N., Chaparadza, R., Prakash, A.: Addressing Stability of Control-Loops in the Context of the GANA Architecture: Synchronization of Actions and Policies. In: Spyropoulos, T., Hummel, K.A. (eds.) IWSOS 2009. LNCS, vol. 5918, pp. 262–268. Springer, Heidelberg (2009)
11. Mortier, R., Kiciman, E.: Autonomic Network Management: Some Pragmatic Considerations. In: INM 2006: Proceedings of the 2006 SIGCOMM Workshop on Internet Network Management, pp. 89–93. ACM, New York (2006)
12. Kastrinogiannis, T., Tcholtchev, N., Prakash, A., Chaparadza, R., Kaldanis, V., Coskun, H., Papavassiliou, S.: Addressing stability in future autonomic networking. In: Pentikousis, K., Agüero, R., García-Arranz, M., Papavassiliou, S. (eds.) MONAMI 2010. LNICST, vol. 68, pp. 50–61. Springer, Heidelberg (2011)

13. Series Z: Languages and General Software Aspects for Telecommunication Systems - Formal description techniques (FDT) - Specification and Description Language (SDL), http://www.itu.int/ITU-T/studygroups/com10/languages/Z.100_1199.pdf
14. Lollini, P., Giandomenico, F.D., Bondavalli, A.: A Modeling Methodology for Hierarchical Control Systems and its Application. Journal of the Brazilian Computer Society 10(0104-6500), 57-69 (2005)
15. MDA Specifications (2009), http://www.omg.org/mda/specs.htm
16. Unified Modeling Language (2010), http://www.uml.org/
17. Green, D., DiCaterino, A.: A Survey of System Development Process Models. Tech. rep., Center for Technology in Government (February 1998), http://www.ctg.albany.edu/publications/reports/survey_of_sysdev
18. Sztipanovits, J., Karsai, G., Franke, H.: Model-Integrated Program Synthesis Environment. In: IEEE International Conference on the Engineering of Computer-Based Systems, p. 348 (1996)
19. Lédeczi, A., Bakay, A., Maróti, M., Völgyesi, P., Nordstrom, G., Sprinkle, J., Karsai, G.: Composing Domain-Specific Design Environments. Computer 34(11), 44–51 (2001)
20. Bakshi, A., Prasanna, V.K., Ledeczi, A.: MILAN: A Model Based Integrated Simulation Framework for Design of Embedded Systems. In: LCTES 2001: Proceedings of the ACM SIGPLAN Workshop on Languages, Compilers and Tools for Embedded Systems, pp. 82–93. ACM, New York (2001)
21. Lédeczi, Á., Bakay, Á., Maróti, M.: Model-Integrated Embedded Systems. In: Robertson, P., Shrobe, H.E., Laddaga, R. (eds.) IWSAS 2000. LNCS, vol. 1936, pp. 99–115. Springer, Heidelberg (2001)
22. Institute for Software Integrated Systems at Vanderbilt University: Generic Modelling Environments, version 7, http://www.isis.vanderbilt.edu/projects/gme/
23. Chaparadza, R., Prakash, A., Theisz, Z., Zafeiropoulos, A., Jaekel, C., Zahemszky, A., Rubio, C.M., Jokikyyny, T.: Model-driven Methodology and associated Tool-Chain for design for Stability in GANA, Project Deliverable - D1.8b (June 2011), www.efipsans.org
24. Jensen, K.: Coloured Petri Nets: Basic Concepts, Analysis Methods and Practical Use. In: Monographs in Theoretical Computer Science, 2nd corrected printing edn. Basic Concepts, vol. 1, Springer, Heidelberg (1997) ISBN: 3-540-60943-1
25. Jensen, K.: Coloured Petri Nets: Basic Concepts, Analysis Methods and Practical Use. In: Monographs in Theoretical Computer Science. Practical Use, vol. 3, Springer, Heidelberg (1997) ISBN: 3-540-62867-3
26. Jensen, K.: Coloured Petri Nets: Basic Concepts, Analysis Methods and Practical Use. In: Monographs in Theoretical Computer Science, 2nd corrected printing edn., vol. 2. Springer, Heidelberg (1997) ISBN: 3-540-58276-2
27. Jensen, K., Rozenberg, G. (eds.): High-level Petri Nets: Theory and Application. Springer, London (1991) ISBN: 3-540-54125 X or 0-387-54125 X
28. Unbehauen, H.: System Identification Methods using Parameter Estimation - A Survey. Annual Review in Automatic Programming 12(Part 1), 69–81 (1985)
29. Walsh, W.E., Tesauro, G., Kephart, J.O., Das, R.: Utility Functions in Autonomic Systems. In: ICAC 2004: Proceedings of the First International Conference on Autonomic Computing, pp. 70–77. IEEE Computer Society, Washington, DC, USA (2004)

30. Rétvári, G., Németh, F., Prakash, A., Chaparadza, R., Hokelek, I., Fecko, M., Wódczak, M., Vidalenc, B.: A Guideline for Realizing the Vision of Autonomic Networking: Implementing Self-Adaptive Routing on top of OSPF. In: Formal and Practical Aspects of Autonomic Computing and Networking: Specification, Development and Verification. IGI Publishing (2010)
31. Coltun, R., Ferguson, D., Moy, J.: OSPF for IPv6. RFC 2740 (Proposed Standard), obsoleted by RFC 5340 (December 1999),
    http://www.ietf.org/rfc/rfc2740.txt
32. OPNET IT Guru Academic Edition,
    http://www.opnet.com/university_program/itguru_academic_edition/
33. MATLAB and Simulink for Technical Computing, http://www.mathworks.com/
34. Eaton, J.W.: GNU Octave Manual. Network Theory Limited (2002)
35. Aldazabal, A., Baily, T., Nanclares, F., Sadovykh, A., Hein, C., Ritter, T.: Automated Model Driven Development Process. In: ECMDA workshop on Model Driven Tool and Process Integration. Fraunhofer IRB Verlag, Stuttgart (2008) ISBN: 978-3-8167-7645-1
36. Hein, C., Ritter, T., Wagner, M.: Model-Driven Tool Integration with ModelBus. In: FTMDD 2009: Proceedings of the First International Workshop on Future Trends of Model-Driven Development (2009), www.modelbus.org
    http://www.modelbus.org/modelbus/images/stories/
    docs/toolint_with_modelbus.pdf
37. Batori, G., Theisz, Z., Asztalos, D.: Domain Specific Modeling Methodology for Reconfigurable Networked Systems. In: Engels, G., Opdyke, B., Schmidt, D.C., Weil, F. (eds.) MODELS 2007. LNCS, vol. 4735, pp. 316–330. Springer, Heidelberg (2007)
38. Bengtsson, J., Larsen, K.G., Larsson, F., Pettersson, P., Yi, W.: UPPAAL — a Tool Suite for Automatic Verification of Real–Time Systems. In: Alur, R., Sontag, E.D., Henzinger, T.A. (eds.) HS 1995. LNCS, vol. 1066, pp. 232–243. Springer, Heidelberg (1996)
39. Jensen, K., Kristensen, L.M., Wells, L.: Coloured Petri Nets and CPN Tools for modelling and validation of concurrent systems. Int. J. Softw. Tools Technol. Transf. 9(3), 213–254 (2007)
40. Microsoft Office Visio 2007 SDK (October 2006),
    http://msdn.microsoft.com/en-us/library/ms409183%28office.12%29.aspx
41. Moorthy, P.N.: Building a Tool for Synthesis of Correct Design from Interaction Specifications. Master's thesis, University of California, San Diego (2006)
42. Varga, A., Hornig, R.: An overview of the omnet++ simulation environment. In: Simutools 2008: Proceedings of the 1st International Conference on Simulation Tools and Techniques for Communications, Networks and Systems & Workshops. ICST (Institute for Computer Sciences, Social-Informatics and Telecommunications Engineering), pp. 1–10. ICST, Brussels (2008)
43. Quagga Routing Suite, http://www.quagga.net/
44. Ferariu, L., Patelli, A.: Migration-based multiobjective genetic programming for nonlinear system identification. In: SACI, pp. 475–480 (2009)
45. EFIPSANS - Exposing the Features in IP version Six protocols that can be exploited/extended for the purposes of designing/building Autonomic Networks and Services: EC FP7-IP Project, INFSO-ICT-215549 (2008-2010),
    www.efipsans.org

Self Organization for Area Coverage Maximization and Energy Conservation in Mobile Ad Hoc Networks

Cem Şafak Şahin[1], M. Ümit Uyar[1,2], Stephen Gundry[2], and Elkin Urrea[1]

[1] The Graduate Center of the City University of New York, New York, NY 10016
[2] The City College of the City University of New York, New York, NY 10031
{csahin,eurrea}@gc.cuny.edu,
{sgundry00,uyar}@ccny.cuny.edu

Abstract. Mobile Ad hoc Networks (MANETs) are widely used for a large number of strategic applications from military to commercial tasks including disaster area discovery, mine field clearing, and transportation systems. In realistic applications, it is not feasible to deploy mobile nodes manually or using a centralized controller. We provide a nature-inspired approach to achieve self-organization of mobile nodes over unknown terrains. In this framework, each mobile node uses a genetic algorithm as a self-distribution mechanism to decide its next speed and movement direction to obtain a uniform distribution. We present a formal analysis of the effectiveness of our genetic algorithm and introduce an inhomogeneous Markov chain model to prove its convergence. The experiment results from our simulation software and our VMware-based testbed show that our nature-inspired algorithm delivers promising results for uniform distribution of mobile nodes over unknown terrains.

Keywords: Genetic Algorithms, Nature Inspired Algorithms, Mobile Ad hoc Networks, Self-organization, Inhomogeneous Markov chain, Topology Control.

1 Introduction

Recently, there has been a growing interest in the self-organization of mobile nodes in Mobile Ad hoc Networks (MANETs) to dynamically adjust their position to maximize the area coverage over an unknown terrain. Mobile nodes in a MANET typically have a small computing capability with a limited memory and power. Consequently topology control algorithms for MANETs must be simple, efficient, easy to adapt for different missions, autonomous, and designed to minimize the energy consumption while maximizing the area coverage.

In this paper, a genetic algorithm (GA) based framework is used to achieve self-organization of mobile nodes. Nature-inspired approaches have proven to be successful in the past for solving different types of problems in ad hoc networks such as swarm behavior [1,2,3], centralized reasoning systems for multi-robot teams [4], ant colony applications [5,6], and robotic communications for collision

M.L. Gavrilova et al. (Eds.): Trans. on Comput. Sci. XV, LNCS 7050, pp. 49–73, 2012.
© Springer-Verlag Berlin Heidelberg 2012

avoidance and map exploration [7,8]. The underlying reasoning is that nature has been shown to be able to successfully deal with scale and complexity issues in a decentralized manner. The goal for self-organizing topology control algorithms is to provide the ability for mobile nodes to organize themselves in a purposeful manner without any manual intervention or a centralized controller. This approach enables a decision-making process to dynamically evolve reconfiguration plans at run time by each mobile node. We provide an autonomous mechanism for each node so that it can make its own movement decisions for a better location based on its local environment (e.g., neighboring nodes and obstacles). Self-organization emerges without the need of external coordination or human intervention as the mobile nodes autonomously adapt their positions on the basis of a local view of the surrounding scenario. Our approach shows the basic self-organizing properties for an autonomous system (i.e., all mobile nodes) such as self-configuration, self-adaptation and self-healing since it does not require (i) global information or a centralized controller, (ii) any synchronization among mobile nodes, and (iii) re-programming of agent software due to malfunction or loss of mobile nodes. Our GA-based software agent running autonomously in each mobile node provides a dynamic deployment capability while improving the area coverage in terms of uniformity and completeness. We introduce a GA, called FGA, inspired by the equilibrium of the molecules in physics where each molecule tries to be in the balanced position and to spend minimum energy to protect its own position. We implemented simulation software and a testbed to evaluate FGA's effectiveness and applicability to real-life problems in the Bio-inspired Computing Laboratory at the City College of New York.

This paper is organized as follows. In Section 2, we review prior research on the use of GAs on mobile node deployment, target localization in ad hoc networks, and different methods for analyzing convergence of stochastic algorithms. In Section 3, we briefly introduce GAs, outline our FGA approach, and prove the convergence of our GA-based approach. The section 4 introduces a Markov model of our FGA and shows the convergence using inhomogeneous Markov chain. Simulation experiment and VMware testbed results are presented in Section 5.

2   Background

There have been a number of proposals for the self-organization of mobile nodes. One approach rely on computational geometry techniques such as Voronoi diagrams (VA) where the area of interest is partitioned into many subareas, one for each node, so that the nodes can move to maximize coverage in its own sub-area. [9] applies VAs to discover areas not covered by any node and provide general rules based on the principle of moving nodes from densely deployed areas to sparsely deployed areas. A modification of this approach [10] analyzes the problem of sensor deployment in a hybrid scenario, with both mobile and fixed sensors in the same environment. [11] uses Delaunay triangulation techniques to maximize coverage area while minimizing coverage gaps and overlaps by adjusting the deployment layout of nodes close to equilateral triangulation. [12]

presents algorithms designed to minimize moving distance of nodes with particular emphasis on operative settings where coverage does not imply connectivity.

Another common approach uses potential-field-based algorithms that assume that the movement of each node can be affected by virtual forces (VF) from other nodes and obstacles [13,14,15]. This approach imitates the behavior of electromagnetic particles: when two electro-magnetic particles are too close to each other, a repulsive force pushes them apart. Other approaches are inspired by the physics as well. Based on fluid dynamics, [16] models the mobile node network as as a fluid body and the individual nodes as fluid elements that can penetrate and diffuse into highly unknown and unstructured terrain. In [17], the theory of gas is used to give a unified solution to the problem of deployment and dynamic relocation of mobile sensors in an open environment. A similar approach is used in [18] to model sensor movements in the presence of obstacles.

Although the cited approaches assume unknown environments, all of them require the manual tuning of constants and thresholds whose appropriate values are closely dependent on the particular operative scenario. Our topology control algorithm, called FGA, uses a genetic algorithm in the decision-making process of adaptive systems to dynamically evolve reconfiguration strategies that balance competing objectives at run time. FGA uses only local available information and does not require manual tuning of key parameters or prior knowledge of the operative scenario.

Markov chains are widely used to analyze stochastic algorithms. In [19], a Markov chain analysis of GAs for a modified elitist strategy is studied. The local convergence of a GA with Cauchy mutation operator is analyzed in [20]. [21] considers the convergence of a GA in which only the crossover operator is used with an elitist selection scheme. [22] discusses various metrics used to characterize digital images and lays out the key problems in analyzing noisy images. It then provides a basis for describing digital images in terms of random fields. Further it provides fundamental convergence analysis techniques laid out by R. Dobrushin. [23] analyzes GAs making use of a property called niching where multiple high fitness peaks are preserved by a niche operator.

GA has also proven as an efficient approach in various distributed robotic applications. In[24], a GA-based approach is used to satisfy a distance-safety criteria for a mobile robot motion. [25] proposes an algorithm to guide autonomous robots in a highway to reach to their destination without any collision. In [26], an adaptive GA is used to identify targets while avoiding obstacles. The mobile robots collect information from the environment with their video camera and light sensors and run their own GA to stay away from static and unknown blockages and arrive to a given target.

Our approach has fundamental differences from the existing research cited above. FGA has the basic self-* properties of autonomous system, i.e. self-healing, self-configuration, and self-adaptation, giving rise to a fully decentralized algorithm. A mobile node running FGA adapts itself to a dynamic environment without the need of external coordination or explicit control. Our approach does not need global network knowledge, it can be used as a real-time controller by each mobile node to decide its next speed and movement direction. Another difference of our FGA is that, it is resilient to mobile node losses. It is also important to note that the self-* is more challenging in MANETs compared to sensor networks since, unlike sensor networks, there are no stationary nodes are present in MANETs.

3   Nature Inspired Algorithm for Coverage Maximization and Energy Conversation

We address a challenging problem which is typical in many civilian and military applications defined as follows:

> *Given N mobile nodes with communication ranges of R_{com}, how should they <u>autonomously</u> deploy themselves in an unknown region of interest (A_{RoI}) using <u>only</u> local neighborhood information so that resulting configuration <u>maximizes</u> the area coverage of the network while reducing the battery usage?*

This problem becomes even more challenging due to typical characteristic of MANETS:

- the deployment cannot be determined *a priori* when A_{RoI} is unknown or hostile (for military missions)
- the geographical area may change dramatically in a short time-span during an operation
- the number of mobile nodes may change (increase or decrease) dynamically
- mobile nodes do not have access to navigation maps nor to GPS devices but can only have limited information from local neighbors
- mobile nodes are typically deployed into the terrain from a single entry point (more difficult to analyze than random or other types of initial distributions often seen in existing research).

Since there is no deterministic solution for this problem, we will use GAs to find acceptable solutions as explained in the remainder of this section.

3.1   Genetic Algorithms

Genetic Algorithms (GAs) are stochastic-based search techniques inspired by biological evolution mechanisms such as *reproduction, mutation, recombination,* and *selection* [27,28]. The basic concept of GAs is to simulate these mechanisms in a natural system necessary for evolution, specifically those that follow the principles of survival of the fittest. GAs have a tendency to succeed in an environment in which there is a very large set of candidate solutions (i.e., *individuals*), and in which the search space is uneven and nonlinear with hills and valleys [29,30,31].

The most general form of GA works like this: a *population* is composed of a group of individuals created randomly. The individuals in the population are then evaluated. The evaluation function is given by the programmer and provides the individuals a score based on how well they perform at the given task.

Two individuals are then selected based on their fitness scores, the higher the fitness then the higher the probability of being selected. These individuals then reproduce to create one or more *offspring*, after which the offspring are mutated randomly. This continues until a suitable solution has been found or a certain number of generations have passed.

GA Operators. GAs mimic basic principles of life and apply genetic operators like selection, crossover (recombination), or mutation to a sequence of alleles. The sequence of alleles is the equivalent of a *chromosome* in nature and is constructed by a representation which assigns a string of symbols to every possible solution of the optimization problem. For selecting highly fit individuals for reproduction a large number of different selection schemes have been developed. The most popular is proportionate selection (i.e., roulette wheel selection) where individuals are given a probability of being selected that is directly proportionate to their fitness. Two individuals are then chosen randomly based on these probabilities and produce offspring.

Crossover produces two new offspring from two individuals by exchanging substring. The two most common forms of crossover in *gas* are one-point and two-point crossover. In one-point crossover, a position in the chromosome is selected at random and the parts of two parents after the crossover position are exchanged. In two-point crossover, two positions are chosen at random and the segments between them are exchanged.

Mutation allows the introduction of new genetic material by slightly changing the genotype of an individual. Mutation is important for local search and helps prevent the GA from getting stuck at a local optimum. The mutation operator iterates over all genes and mutates each with some low probability of μ_m. The probability of mutation μ_m must be selected to be at a low level because otherwise mutation would randomly change too many alleles and the new individual would have nothing in common with its parent.

Chromosome. A *chromosome* is the genetic material of an individual that represents the information about a possible solution to the given problem. In order to speed up a GA process in the mobile nodes, chromosomes should be selected carefully enough to include all the correct parameters and simple enough to reduce the computational processing power. In general, a binary chromosome may be defined in a discrete search space Ω. A GA minimizes a corresponding fitness function $f_i : \Omega \to \Re$. Then, each potential solution in the search space Ω can be coded into l-bit binary chromosome that is an element of $\Theta = \{0,1\}$. In other words, this is an one-to-one mapping $\Psi = \Omega \to \Theta$.

Fitness Function. In GAs, a *fitness function* (i.e., objective function) is used to measure the quality of a chromosome within a solution space. Fitness calculation of each chromosome in a population is the most time consuming process in a GA. Hence, it must be defined carefully to avoid unnecessary calculation. The effectiveness of a fitness function in a GA is measured by the separation quality of fitness result for each chromosome in a search space Ω.

Algorithm 1. Pseudo-code of our FGA

```
while ! (stopCrit) do
    for all all neighbors do
        if Neighbor is in the neighborhood table then
            Update the neighbor's information
        else
            Add the neighbor's information into table
        end if
    end for
    g ← 0 (generation counter)
    Initialize population P(0)
    Evaluate population P(0)
    while ! (evolveDone) do
        g ← g + 1
        Select P(g) from P(g-1)
        Crossover P(g)
        Mutate P(g)
        Evaluate P(g)
        if localPositionFound then
            evolveDone := true
        end if
    end while
    if betterLocationFound then
        Move to new location
    else
        Wait (Δ T)
        Update neighborhood table
    end if
    Update stopCrit
end while
```

3.2   Our Force-Based Genetic Algorithm (FGA)

We introduced a force-based genetic algorithm called FGA [32,33] inspired by the molecular repulsive force-based distribution in physics [34]. FGA is run by each mobile node as a stand alone software agent. A virtual force is assumed to be applied by the neighboring nodes to a corresponding node. At the equilibrium, the aggregate virtual force applied to a node by its neighbors should sum to zero. If the virtual force is not zero, our GA-based agent uses this non-zero virtual force value in its fitness calculation to adjust the node's speed and direction of movement such that the total virtual force on the mobile node will be minimized. The value of this virtual force depends on the number of neighboring nodes within its communication range of R_{com} and the distance among them. In FGA, a smaller fitness value indicates a better position for the corresponding node.

In FGA framework, each node maintains a *neighborhood table* to keep records for its neighboring mobile nodes. Every ΔT time units, a mobile node N_i runs its GA-based software agent to find a better location to move (if exists) based on the information from its neighborhood table; if it cannot find a location to improve its fitness, the node stops moving momentarily (the details of FGA fitness function are presented in Section 3.1).

Algorithm 1 presents the pseudo code of our FGA. First the neighborhood table is updated by the information received from the nodes in its communication range. Then a population of N individuals, called initial population, is randomly generated where each individual represents a speed and movement direction for the node. Each individual (i.e., chromosome) is then evaluated using a fitness score and sorted based on their fitness values. Since our FGA is posed as a minimization problem, individuals are sorted in decreasing order of fitness scores, representing virtual forces applied to them. In selection and crossover operators in Algorithm 1, individuals are paired for breeding purposes. The mating probability is proportional to their fitness scores (this method is called *roulette wheel selection* [27,35,36]). The offspring are added to a pool as candidate solutions for a new population $P(g+1)$ based on the current one $P(g)$. The pool of candidate solutions has both offspring and P(g). After the offspring in the pool are evaluated, only the better performing individuals are accepted into the newly created population of $P(g+1)$. Mutation occurs on randomly selected individuals of a new population to protect the populations against local optimum points. The population evolves using this process for many generations until a termination criterion (locationPositionFound) is satisfied (e.g., convergence tolerance of the best individuals reaches a certain limit, fitness value becomes below a predefined value, or the number of generations exceeds its limit). If FGA evolves to a better speed and movement direction to minimize the total virtual force on the corresponding node, mobile node adapts this new speed and direction; otherwise, it stops. A node repeats running FGA in this manner until the condition called *stopCrit* is satisfied when the node obtains an acceptable level of uniformity in its vicinity.

3.3   Chromosome in Our FGA

In our mobility model [32,37], each mobile node can move into one of six hexagonal directions in A_{RoI}. As an example, let us assume that mobile nodes can move at four different speeds. Six different directions with four speeds can be coded into 5-bit chromosome ($< d_1d_2d_3s_1s_2 >$). The first three bits ($< d_1d_2d_3XX >$) represent hexagonal movement directions ($< 000 >$ representing north, $< 001 >$ northeast, $< 010 >$ southeast, $< 011 >$ south, $< 100 >$ southwest, and $< 101 >$ northwest). The last two bits of the chromosome ($< XXXs_1s_2 >$) are used for defining different speed values ($< 00 >$ for immobile, $< 01 >$ slower speed, $< 10 >$ normal speed, and $< 11 >$ faster speed). The speed implies the number of hexagonal cells that a node can move in a time unit. For example, if our FGA evolve to a chromosome $< 01110 >$ that means that the corresponding mobile node should move three position (normal speed) heading south.

3.4   Genetic Operators and Fitness Function in Our FGA

Our FGA used one-point crossover (see Section 3.1) with the probability of $\mu_c = 0.9$, roulette wheel selection, and one-bit mutation with the probability of $\mu_m = 0.01$ (see Section 3.1) genetic operators. Each autonomous mobile node gathers information about its neighboring environment including mobile nodes and obstacles within its communication range of R_{com}, and then, runs its own GA-based topology control framework to decide its next speed and movement direction. FGA generates new chromosomes representing candidate solutions for the next generation at each iteration (i.e., generation). These candidate solutions are ordered from the lowest fitness score to the highest. The lowest (i.e., the best) fitness value corresponds to the solution representing the least amount of virtual force applied to the corresponding mobile node. After 30 generations, the best candidate solution having the lowest fitness score is adopted as the new speed and movement direction such that the total force on the corresponding mobile node will be lowered.

Our GA-based topology control algorithm for autonomous mobile nodes use a fitness function that is based on the virtual forces applied to a mobile node by its neighboring nodes [33]. The virtual force between two neighboring nodes (N_i and N_j) depends on the distance between them and the number of other nodes within their communication ranges. The virtual force exerted on node N_i by its neighboring node N_j is calculated as:

$$F_{ij} = \begin{cases} F_{max} & \text{if } d_{ij} = 0 \\ \sigma_i \left(d_{th} - d_{ij} \right) & \text{if } 0 < d_{ij} < d_{th} \\ 0 & \text{if } d_{th} \leq d_{ij} \leq R_{com} \end{cases} \quad (1)$$

where d_{ij} is the Euclidean distance between mobile nodes N_i and N_j, d_{th} is the threshold value to define the local neighborhood, and σ_i is the expected node degree (i.e., it is a function of mean node degree [37] and total number of neighbors of N_i) to maximize the area coverage in A_{RoI}. If the corresponding node N_i has k number of nodes within its communication range, our FGA calculates the fitness value of the autonomous node N_i as:

$$minimize \; : \; \sum_{j=1}^{k} F_{ij} = \sum_{j=1}^{k} \sigma_i \left(d_{th} - d_{ij} \right) \text{ for } 0 < d_{ij} \leq d_{th}$$

$$(2)$$

$$subject \; to \; : \; d_{mov} \leq d_{max}$$

where d_{mov} is a result that encoded in each chromosome whereas d_{max} is the maximum allowable distance based on N_i's neighbors that N_i can move in one time unit. Note that $d_{mov} \leq d_{max}$ since d_{max} depends on the positions of the N_i's neighbors at a given time. For example, in a perfect equilibrium (i.e., all neighboring nodes are at distance of R_{com} from N_i), since N_i does not have to move for a better fitness, d_{max} is zero.

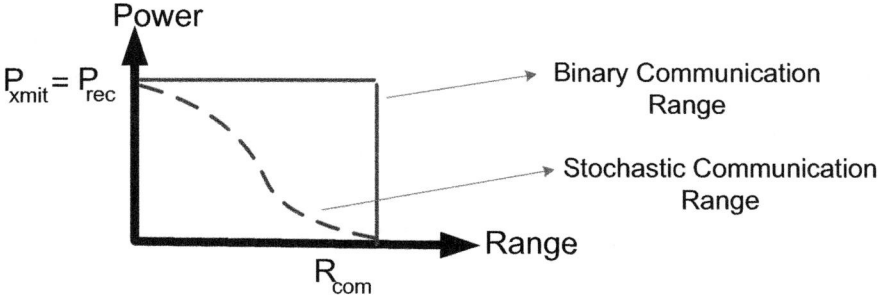

Fig. 1. Binary and stochastic communication range models

Recall that the goal of our GA-based approach is to evolve a chromosome composed of a speed and movement direction that minimize the fitness function (i.e., total virtual force on the corresponding mobile node) given in Eqn. 2. Let us now introduce the following lemma to show the relation between d_{mov} and d_{max}.

Lemma 1. *Our FGA evolves to a distance per time unit d_{mov} such that $0 < d_{mov} < R_{com}$.*

Proof. Based on Eqns. 1 and 2, our FGA generates solutions which cannot result in losing connections between a mobile node N_i and its neighboring nodes. The upper bound of d_{mov} (i.e., the value of d_{max}) is the distance between N_i and its farthest neighbor. Since the maximum distance without losing direct communication is R_{com} we have $0 < d_{mov} < R_{com}$.   □

3.5 Effectiveness in Area Coverage

The concept of area coverage as a function of mobile robots was introduced by Gage [38]. Three types of area coverage are (i) blanket coverage: the objective is to maximize the total occupied area, (ii) barrier coverage: the nodes try to minimize the probability of undetected target penetration thought a barrier, and (iii) sweep coverage: it is equivalent to a moving barrier coverage. We can consider that the dynamic topology control of mobile nodes to maximize the area coverage without global localization information is a combination of blanket and sweep coverage.

Uniform distribution of mobile nodes in a geographical terrain (i.e., a region of interest, A_{RoI}) relates to the issue of how well each point in a node's communication range is covered. One of the main objectives is to deploy mobile nodes in strategic ways (e.g., uniformly) such that a maximum area coverage is achieved according to the needs of the underlying applications. Therefore, effectiveness in area coverage (called *normalized area coverage*, NAC) is an important performance metric for our GA-based topology control framework as it measures the success of mobile nodes' distribution over A_{RoI}.

The communication range of each mobile node can be modeled as either binary or stochastic communication range as seen in Fig. 1. Transmitted power and received power by a mobile node are shown as P_{xmit} and P_{rec} in Fig. 1, respectively. In binary communication range, the probability of communicating with another mobile node within the communication range (R_{com}) is one; otherwise it is zero. However, in the stochastic communication range, the probability of communicating with another mobile node is a decaying function of their distance [14]. In this paper, for simplicity, we prefer using the binary model with the same communication range for all mobile nodes. The results, however, can easily be extended for stochastic communication range model with heterogeneous R_{com} for mobile nodes.

Definition 1. *Suppose a mobile node N_i with a communication range of R_{com} using an omni directional antenna is located at (x_i, y_i). Then a point p at (x_p, y_p) in A_{ROI} is said to be covered by N_i if $\sqrt{(x_i - x_p)^2 + (y_i - y_p)^2} \leq R_{com}$.*

Definition 2. *The effectiveness in the area coverage (called A_{NAC}) is given as:*

$$A_{NAC} = \frac{\bigcup_{i=1}^{n} A_i}{A_{ROI}} \tag{3}$$

where A_i is the area covered by node N_i and $A_{NAC} \in [0, 1]$.

If a node is located well inside the terrain, the full area of a circle around the node with a radius of R_{com} is counted as the covered region if there are no other nodes overlapping with this node's coverage. When the node is near the boundary, only the partial area is included in A_{NAC} computation [39,32].

3.6   Uniformity

Mobile nodes in a MANET have limited energy resources. Uniformly distributed mobile nodes in A_{AoI} use their limited resources more evenly than the non-uniformly distributed nodes for movement and communication.

Definition 3. *Uniformity for a mobile node N_i, called u_i, is defined as the average of the standard deviation of the overlap area of N_i and its neighbors (Fig. 2).*

$$u_i = \sqrt{\frac{1}{m} \sum_{j=1}^{m} (\Lambda_{ij} - \Lambda_{\Theta})^2} \tag{4}$$

where u_i is the local uniformity value for the mobile node of N_i, m is the number of neighbors, Λ_{ij} is the overlap area between N_i and its neighboring node of N_j, and Λ_Θ is the overlap area between N_i and N_j when $d_{ij} = R_{com}$ (if N_i and N_j move any further they will lose communication).

A smaller value of u_i means that neighboring nodes are separated more uniformly. Note that the overlap area between mobile nodes N_i and N_j (see Fig. 3) is $\Lambda_\Theta = R_{com}^2(\Theta - sin(\Theta))$, where $\Theta = 2cos^{-1}(\frac{d_{ij}}{R_{com}})$. The following lemma shows the minimum overlap area between two neighboring nodes.

Lemma 2. *The overlap area between two neighboring nodes N_i and N_j $(1 \leq i, j \leq m)$ is minimized for Λ_Θ where $\Theta = \frac{2\pi}{3}$.*

Proof. If mobile nodes N_i and N_j can communicate with each other, we have $d_{ij} \leq R_{com}$. In order to minimize the overlap area of Λ_{ij}, d_{ij} must be maximized (i.e., $d_{ij} = R_{com}$). As seen in Fig. 3, $\widehat{ABC}$ and $\widehat{ADC}$ are equilateral triangles if $d_{ij} = R_{com}$. Hence, $\Theta = \frac{2\pi}{3}$. $\square$

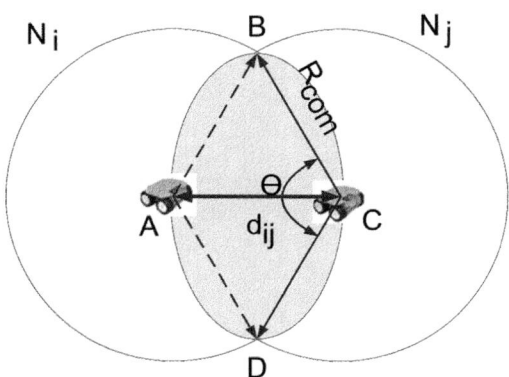

Fig. 2. Overlap area between mobile nodes N_i and N_j when $d_{ij} = R_{com}$

Let us now show that our FGA improves the local uniformity u_i for a mobile node N_i if the fitness value of N_i is greater at time t than at $t + 1$.

Lemma 3. *Our FGA reduces u_i for a mobile node N_i with k neighbors if $f_i^t > f_i^{t+1}$.*

Proof. Using Eqn. 2,

$$f_i^t > f_i^{t+1} \Rightarrow \sum_{j=1}^{k}(\sigma_i(d_{th} - d_{ij}^t)) > \sum_{j=1}^{k}(\sigma_i(d_{th} - d_{ij}^{t+1}))$$

$$\Rightarrow \sum_{j=1}^{k}(d_{ij}^t) < \sum_{j=1}^{k}(d_{ij}^{t+1}) \tag{5}$$

From Eqn. 4 with $\Theta = 2cos^{-1}(\frac{d_{ij}}{R_{com}})$, we have

$$u_i^t = (\frac{1}{m}\sum_{j=1}^{k}(R_{com}^2(2cos^{-1}(\frac{d_{ijt}}{R_{com}}) - sin(2cos^{-1}(\frac{d_{ijt}}{R_{com}}))) - \Lambda_{\frac{\pi}{3}})^2)^{\frac{1}{2}} \tag{6}$$

$$u_i^{t+1} = (\frac{1}{m}\sum_{j=1}^{k}(R_{com}^2(2cos^{-1}(\frac{d_{ijt+1}}{R_{com}}) - sin(2cos^{-1}(\frac{d_{ijt+1}}{R_{com}}))) - \Lambda_{\frac{\pi}{3}})^2)^{\frac{1}{2}} \tag{7}$$

Based on Eqn. 5, $d_{ij}^t > d_{ij}^{t+1}$ which means:

$$\frac{d_{ij}^t}{R_{com}} > \frac{d_{ij}^{t+1}}{R_{com}} \Rightarrow cos^{-1}(\frac{d_{ij}^t}{R_{com}}) < cos^{-1}(\frac{d_{ij}^{t+1}}{R_{com}}) \tag{8}$$

Eqn. 8 yields $u_i^t > u_i^{t+1}$. Since, from Def. 3, a smaller value of u_i implies better uniformity, N_i has better uniformity u_i at time $t+1$.   □

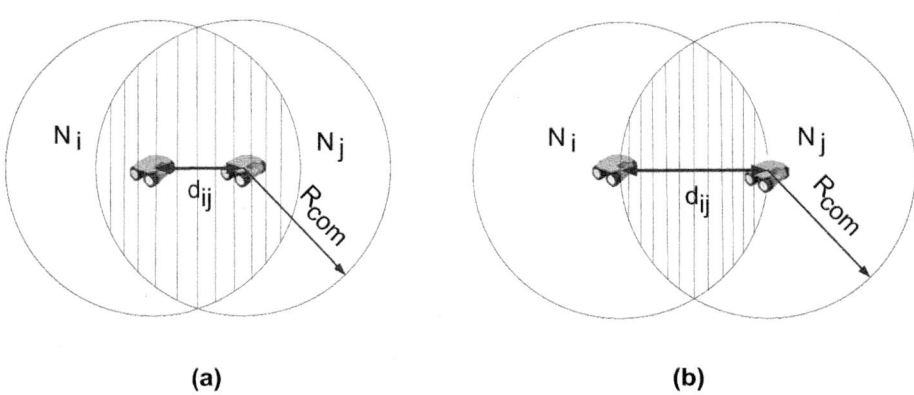

(a)   **(b)**

Fig. 3. Overlap area between mobile nodes N_i and N_j when (a) $d_{ij} < R_{com}$ and (b) $d_{ij} = R_{com}$

Definition 4. *Average of all local uniformities of u_i $(i = 1, \cdots, N)$ yields the uniformity of all mobile nodes as*

$$U = \frac{1}{N} \sum_{i=1}^{N} u_i \tag{9}$$

where N is the total number of mobile nodes in a MANET.

The following lemma states that if there are enough nodes to cover a given a convex terrain, FGA distributes the nodes such that the MANET remains connected.

Lemma 4. *For a convex terrain of A_{RoI} completely covered by mobile nodes N_i and N_j,* FGA *guarantees that N_i and N_j remain connected.*

Proof. Let us consider two conditions contradicting the lemma. First suppose N_i and N_j cover A_{RoI} but do not communicate with each other (i.e., $d_{ij} > R_{com}$) as seen in Fig. 4 (a). From Eqn. 2, since σ_i and σ_j are ∞ when N_i and N_j are not connected, they will start moving to improve their fitness values. FGA will eventually force N_i and N_j to move towards each other such that they will be connected. Note that if N_i and N_j move away from each other more than $2R_{com}$ they no longer cover A_{RoI} and hence contradicted the condition in the lemma.

Now as the second contradictory condition, consider the case that N_i and N_j can communicate with each other (i.e., $d_{ij} < R_{com}$), however, they do not cover A_{RoI} as shown in Fig. 4 (b). From Eqn. 2, FGA will move N_i and N_j to improve their fitnesses by increasing d_{ij} between them in Fig. 4 (c) until the distance is R_{com} which minimizes the local uniformity given in Eqn. 4.

Therefore, FGA guarantees that if N_i and N_j completely cover A_{RoI}, they are connected. □

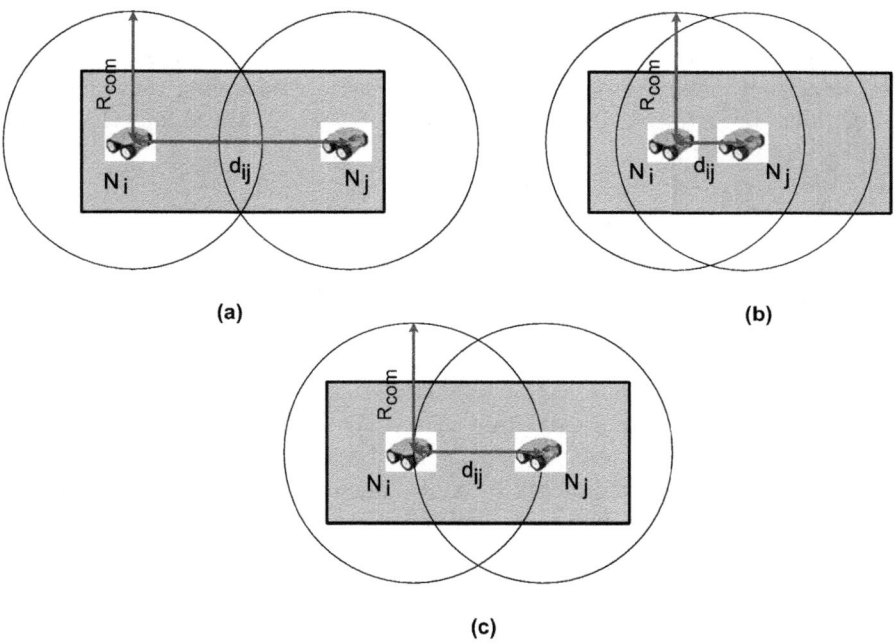

Fig. 4. Area coverage when (a) $d_{ij} > R_{com}$, (b) $d_{ij} < R_{com}$, and (c) $d_{ij} = R_{com}$

4 Markov Chain Model of Our FGA

A *finite Markov chain* is a discrete stochastic process composed of a countable number of random variables referred to as *states* shown as $x \epsilon X$. These states are mutually exclusive where the sum of the probability distributions for all states equals one (i.e., $Pr(X) = Pr(x_1) + \cdots + Pr(x_n) = 1$). Various distributions of the set of random variables are interconnected by means of a transition matrix. If x_{ij} is the probability of transitioning from state i to state j (where $i, j \epsilon X$) in one step, the transition matrix P is given by using x_{ij} as an element in the ith row and j^{th} column:

$$P = \begin{pmatrix} x_{11} & x_{12} & \cdots & x_{1n} \\ x_{21} & x_{22} & \cdots & x_{2n} \\ \cdots & \cdots & \cdots & \cdots \\ x_{n1} & x_{n2} & \cdots & x_{nn} \end{pmatrix}$$

The sum of each row in the transition matrix must be equal to one such that $\sum_j x_{ij} = 1$. For an arbitrary distribution of X denoted as ν to transition to another distribution μ, the transition matrix is applied such that $\nu P = \mu$. It is important to note that the Markov chain exhibits the *memoryless* characteristic where no knowledge of past states is needed to determine the future behavior of the system. It is only important to know the current state of the system and apply the transition matrix to determine future behavior. Assuming that the system transitions through time denoted by X_t (for $t = 1, 2, \cdots$) the memoryless quality can be represented by:

$$Pr(X_t = \mu_t | X_0 = \mu_0, \cdots, X_{n-1} = \mu_{n-1}) = Pr(X_n = \mu_n | X_{n-1} = \mu_{n-1}) \quad (10)$$

Given these characteristics, Markov chains can be used as a method of modeling and analyzing the behavior of our FGA. In our mobility model, the behavior of each node is encompassed in its chromosome representing a fitness, speed and direction and therefore can easily be mapped to distinct states. If each node is observed from one time instant to the next, the change in the chromosome can be used to generate a transition matrix for each corresponding change in the population. Unfortunately, given the number of bits in the chromosome, the system would have approximately 10^6 states due to a fitness resolution of 10^4, 10 different speeds, and six directions. This large number of states quickly becomes computationally unfeasible and makes convergence analysis difficult. A reduced Markov chain model of a mobile agent utilizing our FGA was created by mapping the original state space to a smaller number of states. These states are reduced based on the node's six directions (up, up-right, down-right, down, down-left, or up-left), fitness (good or bad), and speed (mobile or immobile) as shown in Fig. 5. For simplicity, it is assumed that mobile nodes are able to change directions without stopping. The high degree of precision used in calculating fitness while running the FGA helps the mobile node make the best movement decision possible. The goal of the analysis is to determine if the FGA eventually converges to a stable configuration of high fitness. Therefore, the most important aspect of the fitness in the analysis is whether the node has high fitness or not and is correspondingly shown in Fig. 5 as 1 and 0. $d(i)$ is the number of neighbors within R_{com} of node i and $\overline{N}$ is the ideal number of neighbors [37] for maximize the area coverage. The states that fulfill the condition that $(\overline{N}-1) < d(i) < (\overline{N}+1)$ is denoted as *ideal*; and otherwise marked as *non-ideal* (seen as *Non* in Fig. 5). As depicted in Fig. 5, the reduced Markov chain model of our FGA has 15 states. When the mobile node is moving, it is within one of 12 states, relative to the six directions of motion and whether or not it has an ideal number of neighbors. Given that the node is moving in one of the six directions, it can be inferred that the speed is non-zero and that the fitness is non-ideal. One of the three remaining state is (**Stop, Non, 0**) where the node is immobile, while having a non-ideal fitness and number of neighbors. This occurs during initial tightly bound configurations where the node is surrounded and cannot move or the available locations do not improve fitness. The second state is (**Stop, ideal, 0**) where no other location at the present time will increase the nodes fitness; the number of neighbors is ideal, but their configuration is suboptimal for minimal virtual external force. The third state is (**Stop, ideal, 1**) which is the desired state of the algorithm. The node has an ideal number of neighbors with an orientation that is within the optimal threshold fitness using Eqn. 2 and therefore retains its good position until the situation changes and causes the node to once again have suboptimal fitness.

For succinctness Fig. 5 does not show all of the transition edges, but it should be noted, due to the design of our FGA, that every state is connected to every other state.

4.1 Convergence Analysis of Our FGA Modeled as Inhomogeneous Markov Chain

Our FGA has been modeled as a homogeneous Markov chain in [40,39]. Here, this work has been further extended to model the FGA as an inhomogeneous Markov Chain. Inhomogeneous Markov chains are defined as having Markov kernels (i.e., transition matrices) that are different for each time step (i.e., $P_t = P_1, P_2, \cdots$, where $t = 1, 2, \cdots$) over a finite space X, that has some initial distribution ν. The probability distribution over the states $x \epsilon X$ at times $t \geq 0$ is found using $P^{(t)}(x_0, \cdots, x_t) = \nu(x_0)P_1(x_0, x_1) \cdots P_t(x_{t-1}, x_t)$. In [40,39], the homogeneous model clearly shows the differences in the rate of convergence for various experimental parameters (e.g. number of nodes, various values of R_{com}, etc.). The transition matrix of inhomogeneous Markov chain is different at each time unit; therefore, convergence rate can be exercised more accurately than the homogeneous Markov chain model of our FGA explained in [40,39]. Furthermore, inhomogeneous Markov chain also provides the time-based behavior of our FGA with respect to different environmental conditions (e.g., number of neighbors and position of obstacles) rather than average behavior presented in [40,39]. In other words, by having separate Markov kernels for each time step, the inhomogeneous model is benefited by maintaining the precise time-varying experimental data taken during our simulations that clearly demonstrates the model's convergence. These are valuable observations since we can improve our algorithm based on different scenarios.

Before discussing the convergence of various Markov chains, two measures must be defined. First, the total variation between two distributions μ and ν of a finite set of random variables X is defined as $\|\mu - \nu\| = \sum_n |\mu(x) - \nu(x)|$. This is a rough measure of the orthogonality of between any two distributions of a finite set. Using this variation, the Dobrushin's contraction coefficient [22] is defined as $c(P) = \frac{1}{2} \cdot \underset{x,y}{max} |P(x, \cdot) - P(y, \cdot)|$, where c is the contraction coefficient and P is a transition matrix. The contraction coefficient is $\frac{1}{2}$ of the largest total

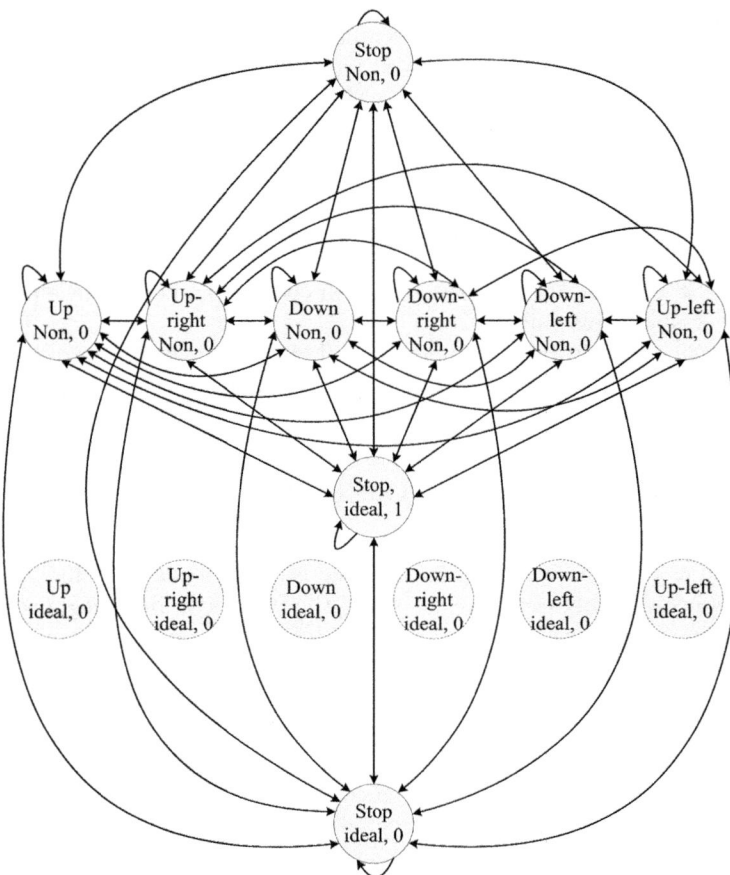

Fig. 5. Markov chain model of our FGA (every state is connected to every other state encircled with dotted lines, whose edges are not shown for succinctness)

variation between all combinations of rows in a transition matrix. Given that any two distributions of the transition matrix are disjoint, then $c(P) = 1$. $c(P) = 0$ when all rows in the transition matrix $P(x, \cdot)$ are equal. Using these measures, the first step required for the convergence analysis is to demonstrate the existence of a limiting distribution by the following lemma taken from [22].

Lemma 5. ([22]) If μ_t, $t \geq 1$, are probability distributions on X such that $\sum_t \|\mu_{t+1} - \mu_t\| < \infty$ then there is a probability distribution μ_∞ such that $\mu_t \to \mu_\infty$ as $t \to \infty$.

Proof. For $s < t$, $\|\mu_t - \mu_s\| \leq \sum_{r \geq s} \|\mu_{r+1} - \mu_r\|$. (See [22] for details.) □

Lemma 5 shows that a limiting distribution μ_∞ exists if any sequence of distributions μ_t exists whose total variations have a finite sum. Assuming that a sequence with a finite sum exists, μ_∞ must also exist within X because it is a finite and closed space. This is commonly referred to as a *Cauchy sequence*. Lemma 5 predicates the following theorem, commonly referred to its founder R. L. Dobrushin:

Theorem 1. *([22]) Let P_t, $t \geq 1$, be transition matrices, each with an invariant probability distribution μ_t. Given that the following conditions exist*

$$\sum_t ||\mu_{t+1} - \mu_t|| < \infty \tag{11}$$

$$\lim_{t \to \infty} c(P_i \cdots P_t) = 0 \ for \ every \ i \geq 1 \tag{12}$$

Then $\mu_\infty = lim_{t \to \infty} \mu_t$ exists and starting from any distribution ν.

$$\nu P_i \cdots P_t \to \mu_\infty \ as \ t \to \infty \tag{13}$$

Proof. Proof is given in [22]. □

Theorem 1 demonstrates that within a finite set X, a given inhomogeneous Markov chain has a limiting distribution as long as two conditions are fulfilled. First, that the sum of the total variation between each time-step's output distribution is finite. Second, the contraction coefficient for the transition matrices eventually go to zero regardless of what point in time i that they begin.

Using Theorem 1 we declare that our FGA converges to a stationary distribution over all states:

Theorem 2. *The set of inhomogeneous transition matrices for FGA fulfills both conditions set by Theorem 1 and, therefore, it will converge to a limiting distribution.*

Proof. (*sketch*) It is shown in [41] that the set of inhomogeneous Markov kernels for our FGA have one step total variations of output distributions whose sum is finite and has a contraction coefficient for the transition matrix that converges to zero from any initial point in time. Therefore, using Theorem 1, our FGA converges to a limiting distribution. □

5 Simulation Software and Testbed Implementation for our FGA

5.1 Simulation Software

In order to study the effectiveness and convergence of our GA-based framework for uniform distribution of autonomous mobile nodes in a terrain, we implemented simulation software in Java [32,37].

Sample screen shots of the graphical user interface of our simulation software are shown in Figs. 6 (a)-(b). As seen in Fig. 6 (a), the total number of nodes (Nodes), communication range of a mobile node (Comm_Range), type of evolutionary algorithms (Case), maximum number of iterations (T_{max}), initial

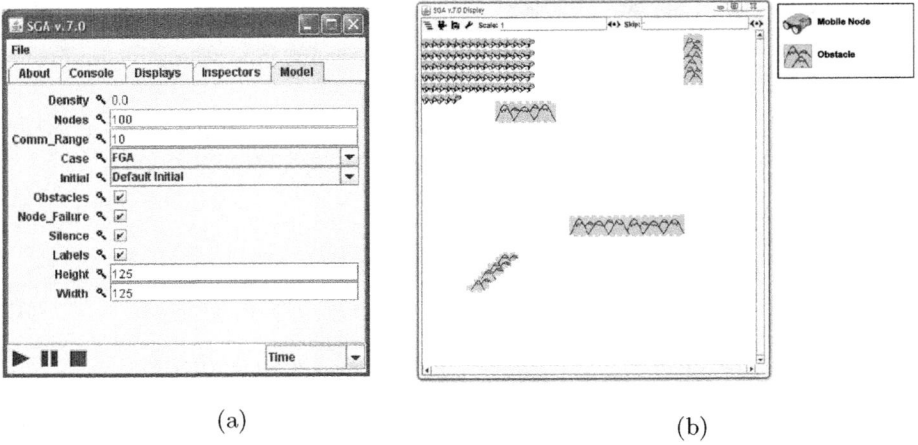

Fig. 6. Graphical user interface for our GA software package (a) a screen shot of user input, and (b) a screen shot from an initial mobile node distribution for FGA

deployment type (Initial), size of A_{RoI} (d_{max}), user defined obstacles, random mobile node failures, and no communication among mobile nodes for a given time (silent mode) are the parameters that the user can define.

Fig. 6 (b) shows a sample initial deployment of autonomous mobile nodes from the northwest sector of A_{RoI}. Currently, our simulation software has three different initial deployment settings for mobile nodes: (i) start from one corner of a terrain, (ii) start from a given coordinate (e.g., the center of the terrain), and (iii) randomly placed in the terrain. A corner initial deployment strategy represents a more realistic approach of the topology control and energy conservation problems for the mobile nodes in an unknown area than other types of deployment strategies.

Note that our simulation software also has the capability of running experiments using a previously used initial mobile node distribution and initial condition from previous runs with the same initial location, speed, and movement direction. This is an important feature, since each experiment is repeated many times to eliminate the noise in collected data, provides an accurate stochastic behavior of our GA-based framework, and allows the building of a Markov chain model for our FGA.

5.2   Convergence Analysis of Our FGA

Experimental Setup. In order to analyze the convergence properties of our FGA using a Markov chain model, we ran experiments for networks with $N = 125$ mobile nodes and $R_{com} = 10$. The mobile nodes deploy in the area of 100x100. All mobile nodes are placed at the northwest of A_{RoI} as shown in Fig. 6 (b). In order to reduce the noise in the collected outcomes and build an efficient reduced Markov model of our FGA, each experiment is repeated 75 times with

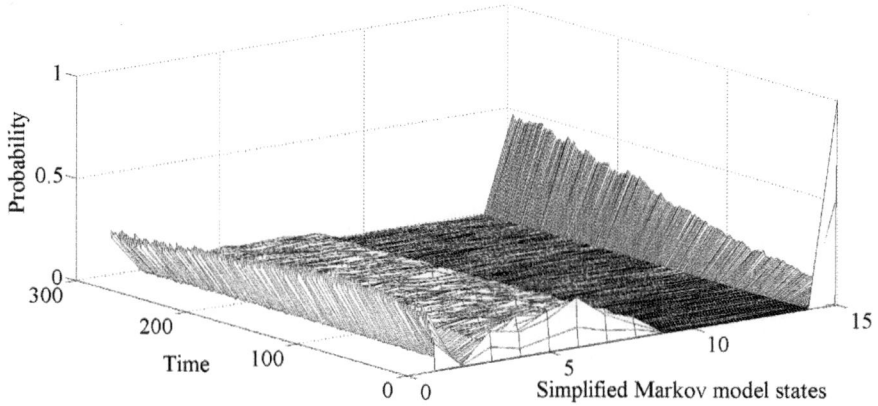

Fig. 7. Distribution of the reduced Markov chain model of our FGA

the same initial values for mobile node speed, movement direction and with the same initial deployment.

Convergence Analysis Using Markov Chain Model. Fig. 7 presents the distribution of the reduced Markov chain model of our FGA and shows that the system evolves to a stationary distribution as time passes. In Fig. 7, the x-axis shows the 15 states of our FGA model from Fig. 5. State 15 is called (Stop, ideal, 1) where the mobile node becomes immobile because it has an ideal number of neighbors with a fitness of 1. The y-axis is the time and the z-axis is the probability of Dobrushin's contraction coefficient. At the beginning of each experiment, most of the mobile nodes are at state (Stop, non, 0) where the mobile node is immobile since all mobile nodes are located at the north-west corner of the geographical terrain (Fig. 6 (b)) and most of them cannot find feasible/available location to move due to the overcrowded region. As seen in Fig. 7, the system evolves to state (Stop, ideal, 1) as time passes. It is important to note, based on Theorem 2, that any initial distribution will converge to the same stationary distribution. The only difference in Fig. 7 will be the convergence time for the different initial distributions. If the mobile nodes are initially distributed in an area such that they are closer to uniform distribution, then they will need less time to achieve uniform distribution. In our experiments, the mobile nodes enter the geographical area from one entry point, which is considered the worst case scenario to reach uniform node distribution. In spite of this, most of the mobile nodes will reach the 15th state of (Stop, ideal, 1).

Fig. 8 shows a rough measure of orthogonality between consecutive instants of time in our reduced Markov chain model (i.e., Dobrushin's contraction coefficient) when time goes to infinity. It is important to note that this figure is only based on the transition matrices and not on various initial distributions of our reduced Markov model.

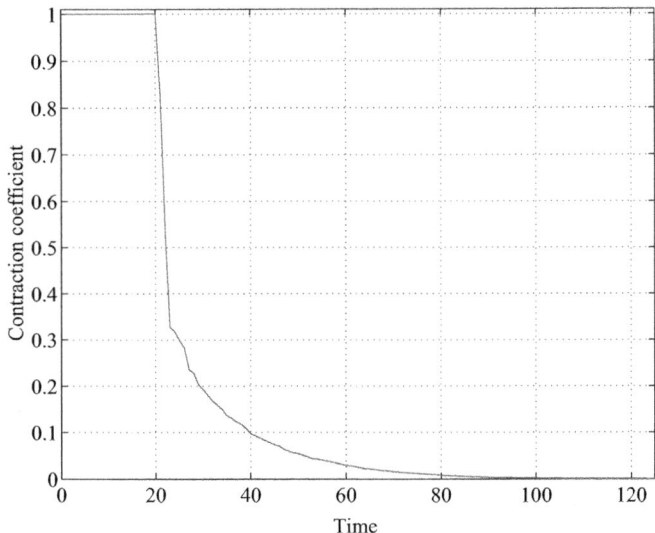

Fig. 8. Contraction coefficients for Markov chain model of our FGA

Therefore, it gives a direct result of the convergence of our FGA and as seen in Fig. 8, the system reaches its final distribution at time $t \approx 100$. Another important observation from Fig. 8, the first 20 time units represent volatility in the system due to the mobile nodes initial placement in one conner of the given terrain. After this period, the nodes have increased degrees of freedom and begin to converge towards to uniform distribution.

The output distribution of each state in our Markov chain model for FGA when time reaches $t = 300$ is shown in Fig. 9. It is important to note that Fig. 9 only presents the output distribution at $t = 300$ since we use inhomogeneous Markov chain model for our FGA. As seen in the figure, the probability of being in state (stop, ideal, 1) has the highest probability of 47%. In other words, at this point nearly half of the mobile nodes have good fitness and are immobile. 23% of the mobile nodes are at state (stop, non, 0). Another important observation from Fig. 9 is that although the system reaches its final distribution at time $t \approx 100$ (as shown in Fig. 8), it converges the final distribution as time passes (i.e., the convergence to the final state is not premature).

Proof of Convergence Using Experimental Data. To study the effectiveness of our FGA we implemented a testbed using VMware virtualization which is an abstraction layer that decouples the physical hardware from the operating system to deliver greater IT resource utilization and flexibility [42]. It is possible to run several virtual machines (VMs) simultaneously on the same physical hardware by isolating each one from the physical environment by using VMware virtualization technique. With this capability, a number of VMs can be implemented on a single real machine to act as mobile nodes. Using this technique

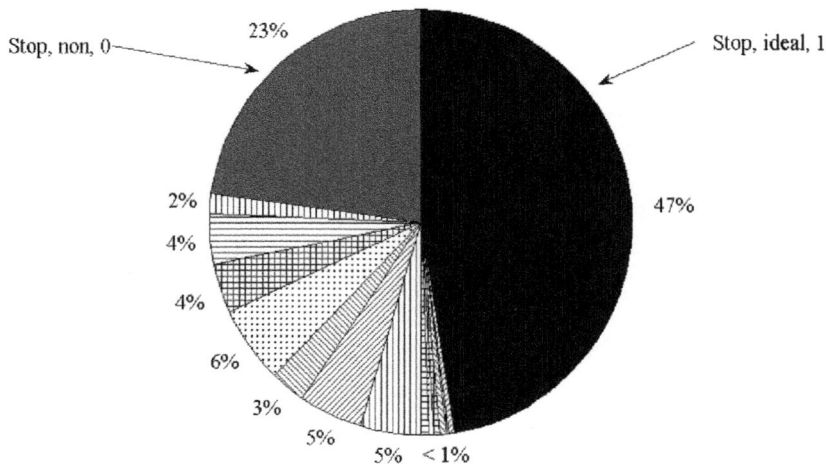

Fig. 9. Output distribution of each state in Markov chain model of our FGA

in our testbed enabled us to scale down the development cost for experiments with large number of mobile nodes and overcoming the limited availability of computer resources/platforms. For simplicity, the mobile nodes in our testbed are configured with the same capabilities, emulating realistic node mobility and wireless features of MANETs including, but not limited to, autonomous mobility, wireless communication characteristics, and periodic heartbeat messages (periodically broadcast to neighboring nodes within R_{com} distance).

Each computer in our testbed has a configurable number of VMs interconnected by a virtual switch, simplifying and allowing our mobile nodes running GA-based framework experimentations on the single computer as though they run on a real network. Typically, a single computer can handle approximately seven nodes for FGA applications. Our testbed implementation is independent of differences between platforms or whether it is actually running on a physical or a virtual machine. This helps to facilitate a flexible deployment paradigm. All VMs are connected to the network through a virtual switch. Typically, all nodes on this network use the TCP/IP protocol suite, although other communication protocols can be used. A host virtual adapter connects the host computer to the private network used for network address translation. Each virtual machine and the host have assigned addresses on the private network. This is done through the DHCP server included with the VMware Workstation.

In our testbed, we ran the experiments for a MANET with $N = 50$ mobile nodes and $R_{com} = 20$ for a total of $T = 300$ time units in a terrain of 100 x 100 units. At the beginning of each experiment, all mobile nodes are located at the northwest corner of the given area (this is similar to the simulation experiment given in Fig. 6 (b)). The mobile node distribution after 300 time units is shown in Fig. 10. We can observe that, in spite of the lack of global knowledge and a centralized controller, the mobile nodes using our FGA obtain an almost uniform coverage of the area in a relatively short period of time. Note that there is

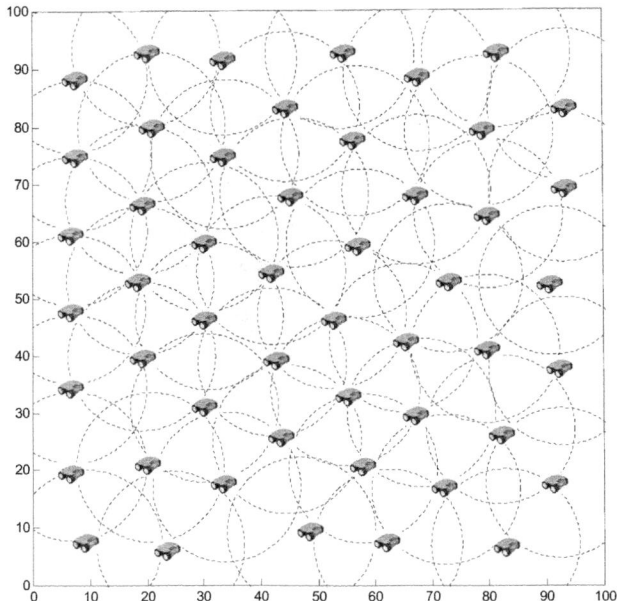

Fig. 10. A screen shot of the final mobile node distribution

no obstacle in this testbed experiment and the mobile nodes can collect more neighborhood information since R_{com} in the testbed experiment is twice of R_{com} used in the simulation experiment.

6   Summary and Conclusion

In this paper, we outlined a nature-inspired topology control approach for efficient self deployment of mobile nodes in a MANET. Our GA-based framework is used by each mobile node as a stand-alone software agent to decide its next speed and movement direction. It does not require any central knowledge and only uses local neighborhood information including the nodes and obstacles in the communication range of a corresponding node. We formally show effectiveness of FGA and present inhomogeneous Markov chain model to prove its convergence. We also provided experiment results from our simulation software and a VMware-based testbed implementation. Analytical and experimental results show that our FGA delivers promising results for uniform mobile node distribution of knowledge sharing nodes over unknown terrains.

Acknowledgments.   This work has been supported by U.S. Army Communications-Electronics RD&E Center. The contents of this document represent the views of the authors and are not necessarily the official views of, or are endorsed by, the U.S. Government, Department of Defense, Department of the Army, or the U.S. Army Communications-Electronics RD&E Center.

This work has been supported by the National Science Foundation grants ECS-0421159 and CNS-0619577.

References

1. Caro, G.D., Ducatelle, F., Gambardella, L.: Swarm intelligence for routing in mobile ad hoc networks. In: Proceedings 2005 IEEE Swarm Intelligence Symposium, SIS 2005, pp. 76–83 (2005)
2. Li, Y., Du, S., Kim, Y.: Robot swarm manet cooperation based on mobile agent. In: 2009 IEEE International Conference on Robotics and Biomimetics (ROBIO), pp. 1416–1420 (2009)
3. Pettinaro, G., Gambardella, L., Ramirez-Serrano, A.: Adaptive distributed fetching and retrieval of goods by a swarm-bot. In: Proceedings of 12th International Conference on Advanced Robotics, ICAR 2005, pp. 825–832 (2005)
4. Franchi, A., Freda, L., Oriolo, G., Vendittelli, M.: A randomized strategy for cooperative robot exploration. In: 2007 IEEE International Conference on Robotics and Automation, pp. 768–774 (2007)
5. Sujatha, B., Harigovindan, V., Namboodiri, M., Sathyanarayana, M.: Performance analysis of pbant (pbant: Position based ant colony routing algorithm for manets). In: 16th IEEE International Conference on Networks, ICON 2008, pp. 1–6 (2008)
6. Wu, Z., Dong, X., Song, H., Jiang, S., Liang, Y.: Ant-based energy aware disjoint multipath routing algorithm in manets. In: 2006 1st International Symposium on Pervasive Computing and Applications, pp. 752–757 (2006)
7. Dai, Y., Hinchey, M., Madhusoodan, M., Rash, J., Zou, X.: A prototype model for self-healing and self-reproduction in swarm robotics system. In: 2nd IEEE International Symposium on Dependable, Autonomic and Secure Computing, pp. 3–10 (September 2006)
8. Zhao, J., Jiang, J., Zang, X.: Cooperative multi-robot map-building based on genetic algorithms. In: 2006 IEEE International Conference on Information Acquisition, pp. 360–364 (2006)
9. Wang, G., Cao, G., Porta, T.F.L.: Movement-assisted sensor deployment. IEEE Transactions on Mobile Computing 5, 640–652 (2006)
10. Wang, G., Cao, G., Porta, T.F.L.: Proxy-based sensor deployment for mobile sensor networks. In: 2004 IEEE International Conference on Mobile Ad-hoc and Sensor Systems, pp. 493–502 (2004)
11. Ma, M., Yang, Y.: Adaptive triangular deployment algorithm for unattended mobile sensor networks. IEEE Transactions on Computers 56(7), 847–946 (2007)
12. Tan, G., Jarvis, S.A., Kermarrec, A.M.: Connectivity-guaranteed and obstacle-adaptive deployment schemes for mobile sensor networks. In: ICDCS 2008: Proceedings of the 2008 The 28th International Conference on Distributed Computing Systems, pp. 429–437. IEEE Computer Society, Washington, DC, USA (2008)
13. Chen, J., Li, S., Sun, Y.: Novel deployment schemes for mobile sensor networks. Sensors 7(11), 2907–2919 (2007)
14. Heo, N., Varshney, P.K.: Energy-efficient deployment of intelligent mobile sensor networks. IEEE Transactions on Systems, Man and Cybernetics, Part A: Systems and Humans 35(1), 78–92 (2005)
15. Zou, Y., Chakrabarty, K.: Sensor deployment and target localization based on virtual forces. In: INFOCOM 2003. Twenty-Second Annual Joint Conference of the IEEE Computer and Communications, March 30, vol. 2, pp. 1293–1303. IEEE Societies (2003)
16. Pac, M.R., Erkmen, A., Erkmen, I.: Towards fluent sensor networks: A scalable and robust self-deployment approach. In: First NASA/ESA Conference on Adaptive Hardware and Systems, AHS 2006, June 15-18, pp. 365–372 (2006)

17. Garetto, M., Gribaudo, M., Chiasserini, C.F., Leonardi, E.: A distributed sensor relocatlon scheme for environmental control. In: IEEE Internatonal Conference on Mobile Adhoc and Sensor Systems, MASS 2007, pp. 1–10 (2007)

18. Kerr, W., Spears, D., Spears, W., Thayer, D.: Two Formal Fluids Models for Multiagent Sweeping and Obstacle Avoidance. In: Hinchey, M.G., Rash, J.L., Truszkowski, W.F., Rouff, C.A. (eds.) FAABS 2004. LNCS (LNAI), vol. 3228, pp. 111–130. Springer, Heidelberg (2004)

19. Suzuki, J.: A markov chain analysis on simple genetic algorithms. IEEE Transactions on Systems, Man, and Cybernetics 4, 655–659 (1995)

20. Rudolph, G.: Local convergence rates of simple evolutionary algorithms with cauchy mutations. IEEE Transactions on Evolutionary Computation 1 (1998)

21. Hong, P., Xing-Hua, W.: The convergence rate estimation of genetic algorithms with elitist. Chinese Science Bulletin, 144–147 (1997)

22. Winkler, G.: Image Analysis, Random Fields and Markov Chains Monte Carlo Methods. Springer, Heidelberg (2006)

23. Horn, J.: Finite markov chain analysis of genetic algorithms with niching. In: Proceedings of the 5th International Conference on Genetic Algorithms, pp. 110–117. Morgan Kaufmann Publishers Inc., San Francisco (1993)

24. Chaiyaratana, N., Zalzala, A.: Recent developments in evolutionary and genetic algorithms: theory and applications. In: Second International Conference on Genetic Algorithms In Engineering Systems:Innovations and Applications (Conf. Publ. No. 446), pp. 270–277 (1997)

25. Shinchi, T., Tabuse, M., Kitazoe, T., Todaka, A.: Khepera robots applied to highway autonomous mobiles. Artif. Life Robotics (2003)

26. Gesu, V., Lenzitti, B., Bosco, G.L., Tegolo, D.: A distributed architecture for autonomous navigation of robots. In: Proc. of the Fifth IEEE Int. Workshop on Computer Architectures for Machine Perception (2000)

27. Holland, J.: Adaptation In Natural and Artificial Systems. University of Michigan Press (1975)

28. Vose, M.D.: The Simple Genetic Algorithm: Foundations and Theory. MIT Press, Cambridge (1998)

29. Yao, J., Kharma, N., Grogono, P.: Bi-objective multipopulation genetic algorithm for multimodal function optimization. Trans. Evol. Comp. 14, 80–102 (2010)

30. Doyle, L.E.: Essentials of cognitive radio, Cambridge, UK. The Cambridge Wireless Essentials Series, pp. 123–154 (2009)

31. Ellabaan, M.M., Ong, Y.S., Lim, M.H., Jer-Lai, K.: Finding multiple first order saddle points using a valley adaptive clearing genetic algorithm. In: Proceedings of the 8th IEEE International Conference on Computational Intelligence in Robotics and Automation CIRA 2009, pp. 457–462. IEEE Press, Piscataway (2009)

32. Sahin, C.S., Urrea, E., Uyar, M.U., Conner, M., Bertoli, G., Pizzo, C.: Design of genetic algorithms for topology control of unmanned vehicles. Special Issue of the Int. J. of Applied Decision Sciences on Decision Support Systems for Unmanned Vehicles 3(3), 221–238 (2010)

33. Sahin, C.S., Urrea, E., Uyar, M.U., Conner, M., Hokelek, I., Bertoli, G., Pizzo, C.: Genetic algorithms for self-spreading nodes in manets. In: GECCO 2008: Proceedings of the 10th Annual Conference on Genetic and Evolutionary Computation, pp. 1141–1142. ACM, New York (2008)

34. Heo, N., Varshney, P.: A distributed self spreading algorithm for mobile wireless sensor networks. IEEE Wireless Communications and Networking (WCNC) 3(1), 1597–1602 (2003)

35. Trianni, V.: Adaptation in Natural and Artificial Systems. Springer, Heidelberg (2008)
36. Mitchell, M.: An Introduction to Genetic Algorithms. MIT Press, Boston (1998)
37. Urrea, E., Sahin, C.S., Uyar, M.U., Conner, M., Hokelek, I., Bertoli, G., Pizzo, C.: Bioinspired topology control for knowledge sharing mobile agents. Ad. Hoc. Netw. 7(4), 677–689 (2009)
38. Gage, D.W.: Command control for many-robot systems. Unmanned Systems Magazine 10(4), 28–34 (1992)
39. Şahin, C.Ş., Gundry, S., Urrea, E., Uyar, M.Ü., Conner, M., Bertoli, G., Pizzo, C.: Markov Chain Models for Genetic Algorithm Based Topology Control in MANETs. In: Di Chio, C., Brabazon, A., Di Caro, G.A., Ebner, M., Farooq, M., Fink, A., Grahl, J., Greenfield, G., Machado, P., O'Neill, M., Tarantino, E., Urquhart, N. (eds.) EvoApplications 2010, Part II. LNCS, vol. 6025, pp. 41–50. Springer, Heidelberg (2010)
40. Sahin, C.S., Gundry, S., Urrea, E., Uyar, M.U., Conner, M., Bertoli, G., Pizzo, C.: Convergence analysis of genetic algorithms for topology control in manets. In: Proceedings of the IEEE Sarnoff Symposium 2010, IEEE (2010) (to appear)
41. Sahin, C.S.: Design and performance analysis of genetic algorithms for topology control problems. The Graduate Center of the City University of New York, Ph.D. Thesis (2010)
42. VMware Inc.: Workstation Users Manual, Workstation 6.5 (2008)

Data Intensive Distributed Computing in Data Aware Self-organizing Networks

Cong-Vinh Phan

Department of IT, NTT University,
298A – 300A Nguyen Tat Thanh Street, Ward 13, District 4, HCM City, Vietnam
pcvinh@ntt.edu.vn
http://www.linkedin.com/in/phanvc

Abstract. Data management in data aware self-organizing networks (DASNs) such as wireless sensor networks (WSNs), mobile ad hoc networks (MANETs) and so on, whose fundamental paradigm is data aware computing, is currently on the spot as one of the priority research areas and research activities are booming recently. At a borderline of changing DASN configuration, data evolve over time. Thus investigating data awareness emerges as a need, on the one hand, for managing data in DASNs, but on the other hand, for modeling, specifying, programming, and verifying DASNs. Moreover, a well-established foundation of data aware computing in such self-organizing networks becomes a requirement. Hence in this paper, we first construct algebraic models for DASNs. Second, algebraic model of data streams is developed and quantitative behaviors for data streams in DASNs are computed. Third, using algebraic models for DASNs we form monoids of data aware self-organizations and shape streams of data aware self-organizations. Finally, a category of data aware self-organizations monoids is established and streams of data aware self-organizations monoids are considered.

Keywords: Data aware self-organizing networks, WSNs, MANETs, Data aware computing, Data streams, Stream calculus, Categorical structures, Category theory, Coinductive principles, Theory of algebras and coalgebras.

1 Introduction

Most of self-organizing networks such as wireless sensor networks (WSNs), mobile ad hoc networks (MANETs) and so on exploit data aware computing as a fundamental paradigm [6]. Such networks are often described as data aware self-organizing networks (DASNs). At a borderline of changing DASN configuration, data evolve over time [21]. Hence data awareness is arising as a potential approach of, on the one hand, data management and observation in DASNs, but on the other hand, modeling, specifying, programming, and verifying DASNs.

The notion of data aware computing has seen a number of developments through various research investigations including both theoretical and practical aspects [32,24,19,18,7,27,28,29,26,30,31,21]. However, achieving a good approach

M.L. Gavrilova et al. (Eds.): Trans. on Comput. Sci. XV, LNCS 7050, pp. 74–107, 2012.
© Springer-Verlag Berlin Heidelberg 2012

for ensuring that data management and observation in DASNs are properly carried out is far from being obvious. In fact, well-established data aware computing in DASNs is now viewed as a topic requiring enhanced research in order to understand and manage well data in DASNs [4,9,24,32,13].

This paper is intended to present a formal computing for data awareness in DASNs on how the data can evolve and the DASNs themselves can be made more evolvable. In other words, as a rigorous approach to solution of the question, our aim is to construct a categorical computing for data awareness in DASNs to tackle this emerging research topic. Especially, categorical data awareness in DASNs is investigated through the following major contents:

- Constructing algebraic models for DASNs.
- Establishing algebraic model of data streams and computing quantitative behaviors for data streams in DASNs.
- Shaping monoids and streams of data aware self-organizations.
- Forming category of data aware self-organizations monoids and streams of data aware self-organizations monoids.

The rest of this paper is organized as follows: The preliminaries related to self-organization, self-organizing networks and categorical terms are presented in section 2. In section 3, algebraic models for DASNs are formed. In section 4, model of data streams in DASNs is developed by categorical approach. Quantitative behaviors of data streams are computed in section 5. In section 6, we construct monoids and streams of data aware self-organizations in DASNs. By the monoids of data aware self-organizations, we establish a category of data aware self-organizations monoids in section 7. Section 8 is a place for presenting streams in the category of data aware self-organizations monoids as monoid homomorphisms. A category of data aware self-organizations monoids is generalized in section 9. Finally, a short summary is given in section 10.

2 Preliminaries

In this section, we recall some existing concepts related to self-organization, self-organizing networks and categorical terms to support our development.

2.1 Self-organization

Self-organization was defined by F. Heylighen as in

> " Self-organization can be defined as the spontaneous creation of a globally coherent pattern out of local interactions " [8].

In other words, in [16], M. Prokopenko said that self-organization is a set of dynamical mechanisms whereby structures appear at the global level of a system from interactions among its lower-level components. Because of its distributed character, self-organization tends to be robust, resisting perturbations [8]. Local interaction and global structure are closely related concepts, the first being the observable phenomenon connected to the second. Local interaction essentially means that behavior of some components of a system may influence the behavior of other components of the system. Global structure specifies the relation between the semantics of components of a system.

In [23], self-organization is generally described as a facet of self-*. Formally, let self-* be the set of self-_s. Each self-_ to be an element in self-* is called a *self-* facet*. That is,

$$\text{self-*} = \{\text{self-\_} \mid \text{self-\_ is a self-* facet}\} \tag{1}$$

Hence, self-organization is in self-*. Moreover, for instance, self-CHOP is composed of four self-* facets of self-configuration, self-healing, self-optimization and self-protection. Therefore, self-CHOP is a subset of self-*. That is, self-CHOP = {self-configuration, self-healing, self-optimization, self-protection} ⊂ self-*. Every self-* facet must satisfy some certain criteria, so-called *self-* properties*. In [35], T.D. Wolf and T. Holvoet classified the self-* properties in autonomic networks.

2.2   Self-organizing Networks

From the notion of self-organization, self-organizing network is considered as a self-organizing system that was defined in

> "*Self-organizing systems work bottom up. They are composed of a large number of components that interact locally according to typically simple rules. The global behavior of the system emerges from these local interactions* " [20].

This definition is concerned with three major factors of self-organizing system:

- *Bottom up*: Each self-organizing system is composed of a large number of components that interact locally according to typically simple rules.
- *Decentralization*: The global behavior of each self-organizing system emerges from their local interactions.
- *Nondeterminism*: The local interactions are random so multiple different global behaviors are possible, without any specification of which the system will be taken.

Consequently, such self-organizing systems, and thus self-organizing networks, must be capable of predicting and controlling themselves the resulting global behavior in order to minimize the need of direct human interventions.

The topic of self-organization has been intensely studied by various areas of engineering including artificial intelligence, control systems, human orientated systems [33,11,14,34,24] and reconfigurable computing systems [22,29,31,32,21]. Significantly, self-organization has been set as an important requirement for new generation distributed networking infrastructures like wireless sensor networks (WSNs), mobile ad hoc networks (MANETs) and so on [15,6,12,36,5,25]. Such networks pose new challenges for supporting data distribution and management on the self-organizing networking infrastructures characterized by decentralization and nondeterminism.

2.3   Categorical Terms

In categorical language [2,3,1], there are some significant concepts which we recall in this section.

Some Basic Terms. Category **Cat** can be viewed as a graph $(Obj(\mathbf{Cat}), Arc(\mathbf{Cat}), s, t)$, where

- $Obj(\mathbf{Cat})$ is the set of nodes we call *objects*,
- $Arc(\mathbf{Cat})$ is the set of edges we call *morphisms* and
- $s, t : Arc(\mathbf{Cat}) \longrightarrow Obj(\mathbf{Cat})$ are two maps called *source* and *target*, respectively.

We write $f : \mathcal{A} \longrightarrow \mathcal{B}$ when f is in $Arc(\mathbf{Cat})$ and $s(f) = \mathcal{A}$ and $t(f) = \mathcal{B}$.

Associated with each object $\mathcal{A}$ in $Obj(\mathbf{Cat})$, there is a morphism $1_{\mathcal{A}} = \mathcal{A} \longrightarrow \mathcal{A}$, called the *identity* morphism on $\mathcal{A}$, and to each pair of morphisms $f : \mathcal{A} \longrightarrow \mathcal{B}$ and $g : \mathcal{B} \longrightarrow \mathcal{C}$, there is an associated morphism $f; g : \mathcal{A} \longrightarrow \mathcal{C}$, called the *composition* of f with g.

The following equations must hold for all objects $\mathcal{A}$, $\mathcal{B}$ and $\mathcal{C}$ and morphisms $f : \mathcal{A} \longrightarrow \mathcal{B}$, $g : \mathcal{B} \longrightarrow \mathcal{C}$ and $h : \mathcal{C} \longrightarrow \mathcal{D}$:

$$\text{Associativity:} \qquad (f; g); h = f; (g; h) \tag{2}$$

$$\text{Identity:} \qquad 1_{\mathcal{A}}; f = f = f; 1_{\mathcal{B}} \tag{3}$$

A morphism $f : \mathcal{A} \longrightarrow \mathcal{B}$ in the category **Cat** is an *isomorphism* if there exists a morphism $g : \mathcal{B} \longrightarrow \mathcal{A}$ in that category such that $f; g = 1_{\mathcal{A}}$ and $g; f = 1_{\mathcal{B}}$.

That is, if the following diagram commutes.

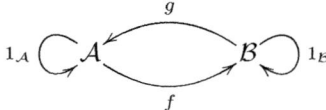

For any set A, $x \in A$ iff $1 \xrightarrow{x} A$ where 1 denotes a singleton set. By the way, we can write $1 \xrightarrow{2} \mathbb{N}$ for $2 \in \mathbb{N}$, $1 \xrightarrow{i} \mathbb{N}$ for $i \in \mathbb{N}$ and so on.

T-Algebra. *Functor* is a special type of mapping between categories. Functor from a category to itself is called an *endofunctor*.

Let **Cat** be a category, $\mathcal{A}$ an object in $Obj(\mathbf{Cat})$, $\mathsf{T} : \mathbf{Cat} \longrightarrow \mathbf{Cat}$ an endofunctor and f a morphism $\mathsf{T}(\mathcal{A}) \xrightarrow{f} \mathcal{A}$; then T-*algebra* is a pair $\langle \mathcal{A}, f \rangle$.

$Obj(\mathbf{Cat})$ is called a *carrier* of the algebra and T a *signature* of the algebra.

T-*homomorphism* between two T-algebras $\langle \mathcal{A}, f \rangle$ and $\langle \mathcal{B}, g \rangle$ is a morphism $\mathcal{A} \xrightarrow{\psi} \mathcal{B}$ such that

$$f; \psi = \mathsf{T}(\psi); g \tag{4}$$

It is equivalent to saying that the following diagram commutes

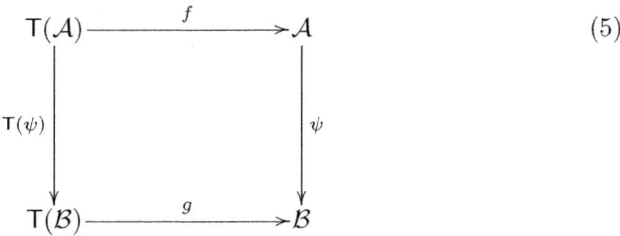

$$(5)$$

3   Algebraic Models for DASNs

DASNs we want to model are intuitionally multiple partial function applications, such as

$$d_0 \xrightarrow{c_0|x_0} d_1 \xrightarrow{c_1|x_1} d_2 \xrightarrow{c_2|x_2} d_3 \ldots \tag{6}$$

where

- All indexes $i \in T\ (= \mathbb{N} \cup \{0\})$ refer to times,
- d is a *data* object in the set, denoted by σ, of data. d_i is the data d at the time i, which makes change of network configuration c_{i-1} to become c_i.
- x_i is a real number that can be thought of as the *multiplicity* (or *weight*) with which the reorganization from c_{i-1} to c_i occurs.
- c is a *network configuration* in the set, denoted by C, of network configurations. c_i is the network configuration c at the time i, which produces data d_{i+1}.

Some first steps of the self-organization process in (6) can also be descriptively drawn as

$$(7)$$

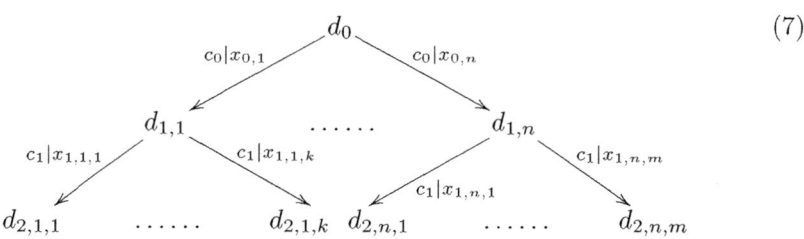

Diagram (7) is thought of as

- For the first step,
  $d_1 \in \{d_{1,1}, \ldots, d_{1,n}\} \subset \sigma$
  and
  $x_0 \in \{x_{0,1}, \ldots, x_{0,n}\} \subset \mathbb{R}$
- For the second step,
  $d_2 \in \{d_{2,1,1}, \ldots, d_{2,1,k}\} \cup \ldots \cup \{d_{2,n,1}, \ldots, d_{2,n,m}\} \subset \sigma$
  and
  $x_1 \in \{x_{1,1,1}, \ldots, x_{1,1,k}\} \cup \ldots \cup \{x_{1,n,1}, \ldots, x_{1,n,m}\} \subset \mathbb{R}$
- and so on

Self-organization process of DASNs that is pictorially drawn in diagram (6) can be separated into two complementary parts as follows:

$$d_0 \xrightarrow{c_0} d_1 \xrightarrow{c_1} d_2 \xrightarrow{c_2} d_3 \ldots \tag{8}$$

and

$$d_0 \xrightarrow{x_0} d_1 \xrightarrow{x_1} d_2 \xrightarrow{x_2} d_3 \ldots \tag{9}$$

On the one hand, diagram (8) emphasizes $c_i \in C$ in the process of producing data streams as formally established in section 4. On the other hand, diagram (9) gives rise to $x_i \in \mathbb{R}$ as weight of c_i in such process of producing data streams and this allows us to evaluate weight-based quantitative behaviors of the data streams as symbolically computed in section 5.

The reorganization process in (8) describes the notion of self-organization on DASNs including the reorganization steps to change DASN *configurations*.

Definition 1. *We define a* DASN *configuration at a reorganization step to be a member of the set* $\sigma^{i \in T} \times C$, *where* $\sigma^{i \in T}$ *stands for*

$$\sigma^{i \in T} = \underbrace{\sigma \times \sigma \times \ldots \times \sigma}_{i \; times} \tag{10}$$

The self-organization paradigm, which we want to approach to, is based on mapping a DASN configuration to another.

Definition 2. *Self-organization on* DASNs *is defined by the following morphism*

$$next : (\sigma^{i \in T} \times C) \longrightarrow ((\sigma^{j \in T} \times C) \longrightarrow \mathbb{R}) \tag{11}$$

The reorganization morphism *next* is nondeterministic. This can be explained as follows: *next* assigns to each DASN configuration in $\sigma^{i \in T} \times C$ a morphism $(\sigma^{j \in T} \times C) \longrightarrow \mathbb{R}$ that can be seen as a kind of *nondeterministic* DASN *configuration* (or so-called *distributed* DASN *configuration*) and specifies for any DASN configuration $\sigma^{j \in T} \times C$ a multiplicity (or weight) $next(\sigma^{i \in T} \times C)(\sigma^{j \in T} \times C)$ in $\mathbb{R}$.

Hence, for example, the meaning of (6) is viewed as the following morphism:

$$next : \sigma \times C \longrightarrow (\sigma \times C \longrightarrow \mathbb{R})$$

(i.e., $next : \sigma^1 \times C \longrightarrow (\sigma^1 \times C \longrightarrow \mathbb{R})$ or denoted by $next^{(\sigma \times C, \sigma \times C \longrightarrow \mathbb{R})}$)

another specific DASN can be specified by

$$next : \sigma \times C \longrightarrow (C \longrightarrow \mathbb{R})$$

(i.e., $next : \sigma^1 \times C \longrightarrow (\sigma^0 \times C \longrightarrow \mathbb{R})$ or denoted by $next^{(\sigma \times C, C \longrightarrow \mathbb{R})}$)

again, we can also specify another specific DASN as

$$next : \sigma^n \times C \longrightarrow (\sigma \times C \longrightarrow \mathbb{R})$$

(i.e., $next : \sigma^n \times C \longrightarrow (\sigma^1 \times C \longrightarrow \mathbb{R})$ or denoted by $next^{(\sigma^n \times C, \sigma \times C \longrightarrow \mathbb{R})}$)

and we can in the completely same way do for any other specific DASN.

The above-mentioned maps, namely *next*, are called *data aware self-organization relations*.

We obtain

Theorem 1. *The sets of DASN configurations as in definition 1 specify a category.*

Proof. In fact, let **Cat(DASN)** be such a category of the sets of DASN configurations, whose structure is constructed as follows:

- Each set $\sigma^i \times C$ of DASN configurations defines an object.
  That is, $Obj(\mathbf{Cat(DASN)}) = \{\sigma^i \times C, \forall i \in T\}$.
- Each *next* defines a morphism.
  That is, $Arc(\mathbf{Cat(DASN)}) = \{next : \sigma^i \times C \longrightarrow \sigma^j \times C\}$.

It is easy to check that associativity in (2) and identity in (3) on all *next*s are satisfied.   Q.E.D.

Theorem 2. *The category* **Cat(DASN)** *equipped with structure* $\sigma^i \times C \longrightarrow ((\sigma^j \times C) \longrightarrow \mathbb{R})$ *defines a category* $\mathbf{Cat(DASN_{\mathbb{R}})}$ *of the sets of DASN configurations.*

Proof. This result comes immediately from theorem 1.   Q.E.D.

Theorem 3. *Structure* $\sigma^i \times C \longrightarrow ((\sigma^j \times C) \longrightarrow \mathbb{R})$ *in category* $\mathbf{Cat(DASN_{\mathbb{R}})}$ *defines an algebra, so-called next-algebra(DASN).*

Proof. This originates from definition on T-algebra in subsection 2.3, where functor T is defined such that $T = next$ with *next* defined in (11).   Q.E.D.

With this result, we obtain a compact formal definition of DASNs as in

Definition 3. *Each next-algebra(DASN) defines a DASN.*

4 Data Streams

Roughly speaking, data stream can be understood as a rope on which we hang up a sequence of data for display. For denoting data stream, the common notation is used as in

$$ds = (d_0, d_1, d_2, \ldots) \tag{12}$$

which specifies a data stream ds indexed by the natural numbers. We can also use the following notation for data stream.

$$ds = (d_i) \qquad \text{for all } i \text{ in } T \tag{13}$$

Hence it follows that

Definition 4. *For morphisms* $1 \xrightarrow{c_i} C$ *and* $1 \xrightarrow{d_i} \sigma$, *there exists a unique morphism* $C \xrightarrow{ds} \sigma$ *such that the equation* $c_i; ds = d_i$ *holds. This is described by the following commutative diagram*

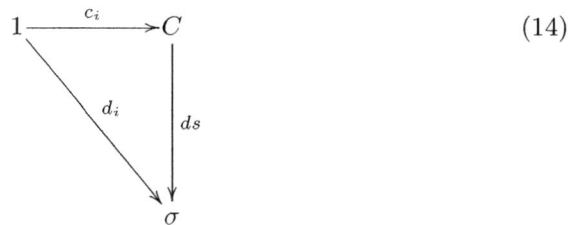

$$\tag{14}$$

Morphism $C \xrightarrow{ds} \sigma$ *defines a data stream.*

Note that morphism $C \xrightarrow{ds} \sigma$ is thought of as

$$\forall c_i [c_i \in C \implies \exists! \, d_i [d_i \in \sigma \;\&\; ds(c_i) = d_i]]$$

In other words, $C \xrightarrow{ds} \sigma$ generates data stream as an infinite sequence of $ds(c_0) = d_0$, $ds(c_1) = d_1$, ..., $ds(c_i) = d_i$, ... which is written as $(ds(c_0), ds(c_1), \ldots, ds(c_i), \ldots)$ or $(d_0, d_1, \ldots, d_i, \ldots)$.

Definition 5. *Given* $C \xrightarrow{ds} \sigma$ *then the set of data streams, denoted by* σ^ω, *is defined by*

$$\sigma^\omega = \{ ds \mid C \xrightarrow{ds} \sigma \} \tag{15}$$

The following result stems immediately from definitions 4 and 5

$$\frac{C \xrightarrow{ds} \sigma}{1 \xrightarrow{ds} \sigma^\omega} \tag{16}$$

Rule (16) means that for each morphism $C\xrightarrow{ds}\sigma$, there is a morphism $1\xrightarrow{ds}\sigma^\omega$ generating member in σ^ω. That is, morphism $C\xrightarrow{ds}\sigma$ generates data streams and $1\xrightarrow{ds}\sigma^\omega$ constructs the set of data streams.

For data streams we define a mechanism to generate them. This mechanism consists of an object C equipping with structural morphisms $1\xrightarrow{c_0}C\xrightarrow{succ}C$ with the property that for σ and any $1\xrightarrow{\sigma_0}\sigma$, $\sigma\xrightarrow{next\_up}\sigma$ there exists a unique morphism $C\xrightarrow{ds}\sigma$ such that the following diagram commutes

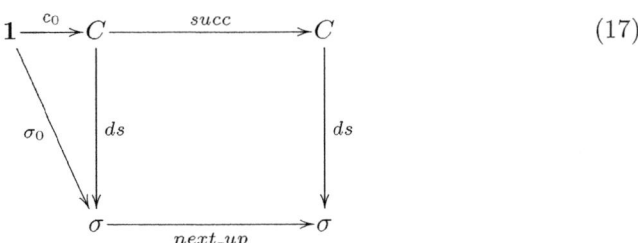

$$(17)$$

Definition 6. *We define a construction morphism of data streams, denoted by* $\bowtie$, *such that*

$$\sigma \times [C\xrightarrow{ds}\sigma]\xrightarrow{\bowtie}[C\xrightarrow{ds}\sigma]$$

This means that

$$\bowtie (A \times B\xrightarrow{f\times g}C \times D) = A \bowtie B\xrightarrow{f\bowtie g}C \bowtie D$$

It follows that any data stream $C\xrightarrow{ds}\sigma$ can be represented in a format including two parts of *head* and *tail* that are connected by $\bowtie$ such that

$$C\xrightarrow{ds}\sigma \equiv \overset{d_0}{\overbrace{1\xrightarrow{c_0}C\xrightarrow{ds}\sigma}} \bowtie \overset{d_{i>0}}{\overbrace{1\xrightarrow{c_{i>0}}C\xrightarrow{ds}\sigma}}$$

where $\overset{d_0}{\overbrace{1\xrightarrow{c_0}C\xrightarrow{ds}\sigma}} = ds(c_0)$ and $\overset{d_{i>0}}{\overbrace{1\xrightarrow{c_{i>0}}C\xrightarrow{ds}\sigma}} = (ds(c_1), ds(c_2),\ldots)$ are called head and tail, respectively.

Definition 7. *We define a head construction morphism, denoted by* $Head(\_)$, *such that*

$$Head(\_) : [C\xrightarrow{ds}\sigma] \longrightarrow \sigma$$

This states that $\forall(a \bowtie s)[(a \bowtie s) \in [C\xrightarrow{ds}\sigma] \implies \exists! \, d_0[d_0 \in \sigma \ \& \ Head(a \bowtie s) = a = d_0]]$.

It follows that $Head(C\xrightarrow{ds}\sigma) \equiv 1\xrightarrow{c_0}C\xrightarrow{ds}\sigma$.

Definition 8. *We define a tail construction morphism, denoted by $Tail(\_)$, such that*

$$Tail(\_) : [C \xrightarrow{ds} \sigma] \longrightarrow [C \xrightarrow{ds} \sigma]$$

This means that $\forall (a \bowtie s)[(a \bowtie s) \in [C \xrightarrow{ds} \sigma] \implies \exists! \langle d_1, d_2, \ldots \rangle [\langle d_1, d_2, \ldots \rangle \in [C \xrightarrow{ds} \sigma]$ & $Tail(a \bowtie s) = s = \langle d_1, d_2, \ldots \rangle]]$.

As a convention, $Tail^{\langle n \rangle}(\_)$ denotes applying recursively the $Tail(\_)$ n times. Thus, specifically, $Tail^{\langle 2 \rangle}(\_), Tail^{\langle 1 \rangle}(\_)$ and $Tail^{\langle 0 \rangle}(\_)$ stand for $Tail(Tail(\_))$, $Tail(\_)$ and $(\_)$, respectively.

It follows that the first member of data stream $C \xrightarrow{ds} \sigma$ is given by

$$Head(Tail(C \xrightarrow{ds} \sigma)) \equiv 1 \xrightarrow{c_1} C \xrightarrow{ds} \sigma$$

and, in general, for every $c_k \in C$ the k-th member of data stream $C \xrightarrow{ds} \sigma$ is provided by

$$Head(Tail^{\langle k \rangle}(C \xrightarrow{ds} \sigma)) \equiv 1 \xrightarrow{c_k} C \xrightarrow{ds} \sigma \tag{18}$$

Data stream to be an infinite sequence of all d_i is viewed and treated as single mathematical entity, so the derivative of data stream $C \xrightarrow{ds} \sigma$ is given by $Tail(C \xrightarrow{ds} \sigma)$.

Now using this notation for derivative of data stream, we can specify data stream $C \xrightarrow{ds} \sigma$ as

- Initial value: $1 \xrightarrow{c_0} C \xrightarrow{ds} \sigma$ and
- Differential equation:

$$Tail(Tail^{\langle n \rangle}(C \xrightarrow{ds} \sigma)) = Tail^{\langle n+1 \rangle}(C \xrightarrow{ds} \sigma)$$

The initial value of $C \xrightarrow{ds} \sigma$ is defined as its first element $1 \xrightarrow{c_0} C \xrightarrow{ds} \sigma$, and the derivative of data stream, denoted by $Tail(C \xrightarrow{ds} \sigma)$, is defined by $Tail(Tail^{\langle n \rangle}(C \xrightarrow{ds} \sigma)) = Tail^{\langle n+1 \rangle}(C \xrightarrow{ds} \sigma)$, for any integer $n \geqslant 0$. In other words, the initial value and derivative equal the head and tail of $C \xrightarrow{ds} \sigma$, respectively. The behavior of a data stream $C \xrightarrow{ds} \sigma$ consists of two aspects: it allows for the observation of its initial value $1 \xrightarrow{c_0} C \xrightarrow{ds} \sigma$; and it can make an evolution to the new data stream $Tail(C \xrightarrow{ds} \sigma)$, consisting of the original data stream from which the first element has been removed. The initial value of $Tail(C \xrightarrow{ds} \sigma)$, which is $Head(Tail(C \xrightarrow{ds} \sigma)) = 1 \xrightarrow{c_1} C \xrightarrow{ds} \sigma$ can in its turn be observed, but note that we have to move from $C \xrightarrow{ds} \sigma$ to $Tail(C \xrightarrow{ds} \sigma)$ first in order to do so. Now a behavioral differential equation defines a data stream by specifying its initial value together with a description of its derivative, which tells us how to continue.

Every member d_i in σ can be considered as a data stream in the following manner. For every d_i in σ, a unique data stream is defined by morphism f:

$$
\begin{array}{ccc}
 & (d_i,\circ,\circ,\dots) & \\
 & \overset{\frown}{\qquad\qquad} & \\
1 \xrightarrow{\quad d_i \quad} \sigma \xrightarrow{\quad f \quad} \sigma^\omega
\end{array}
$$

Such that the following equation holds

$$d_i; f = (d_i, \circ, \circ, \dots)$$

with $\circ$ denoting empty member (or null member) in σ and $(d_i, \circ, \circ, \dots)$ in σ^ω.

Definition 9. *For any $C \xrightarrow{ds1} \sigma$ and $C \xrightarrow{ds2} \sigma$, we have*

$$ds1 = ds2 \text{ iff } 1 \xrightarrow{c_n} C \xrightarrow{ds1} \sigma = 1 \xrightarrow{c_n} C \xrightarrow{ds2} \sigma \tag{19}$$

with every c_n in C.

Definition 10. *Bisimulation on σ^ω is a relation, denoted by $\sim$, between data streams $C \xrightarrow{ds1} \sigma$ and $C \xrightarrow{ds2} \sigma$ such that*

$$
\begin{array}{c}
ds1 \sim ds2 \\
Head(ds1) = Head(ds2) \text{ and } Tail(ds1) \sim Tail(ds2)
\end{array}
\tag{20}
$$

For validating whether $ds1 = ds2$, a powerful method is so-called *proof principle of coinduction* [18] that states as follows:

Theorem 4. *For any $C \xrightarrow{ds1} \sigma$ and $C \xrightarrow{ds2} \sigma$, we have*

$$\frac{ds1 \sim ds2}{ds1 = ds2} \tag{21}$$

Proof. Hence in order to prove the equality of two data streams $ds1$ and $ds2$, it is sufficient to establish the existence of a bisimulation relation $ds1 \sim ds2$. In fact, for two data streams $ds1$ and $ds2$ and a bisimulation $ds1 \sim ds2$. We have
    By induction on $n \geqslant 0$ and bisimulation $\sim$
$\vdash Tail^{\langle n \rangle}(ds1) = Tail^{\langle n \rangle}(ds2)$
    By bisimulation $\sim$
$\vdash Head(Tail^{\langle n \rangle}(ds1)) = Head(Tail^{\langle n \rangle}(ds2))$
    By identity in (18)
$\vdash 1 \xrightarrow{c_n} ds1 = 1 \xrightarrow{c_n} ds2$ with every $c_n \in C$

It follows that, by (19), $ds1 = ds2$. Note that the notation $\vdash$ denotes an inference. By this way, when we write $a \vdash b$, for example, which means that b is true because of a that we have. Moreover, when we write $a \vdash b \vdash c \vdash \dots$ that is understood as a chain inference by which a justification happens.   Q.E.D.

As a consequence, using coinduction we can establish the validity of the equivalence between data streams $C\xrightarrow{ds1}\sigma$ and $C\xrightarrow{ds2}\sigma$ in σ^ω.

Theorem 5. *For all ds in σ^ω,*

$$ds = Head(ds) \bowtie Tail(ds) \tag{22}$$

Proof. This stems from the coinductive proof principle in (21). In fact, it is easy to check the following bisimulation $ds \sim Head(ds) \bowtie Tail(ds)$. It follows that $ds = Head(ds) \bowtie Tail(ds)$.   Q.E.D.

In theorem 5, operation $\bowtie$ as a kind of data stream integration, the theorem states that data stream derivation and data stream integration are inverse operations. It gives a way to obtain ds from $Tail(ds)$ and the initial value $Head(ds)$. Therefore, the theorem allows us to reach solution of differential equations in an algebraic manner.

5   Quantitative Behaviors of Data Streams

For quantitative behaviors of data streams, we consider again the diagram (9) in section 3. Let $B(\_)$ be quantitative behavior of data stream in DASNs.

Definition 11. *For morphisms $1\xrightarrow{c_i}C$ and $1\xrightarrow{x_i}\mathbb{R}$, there exists a unique morphism $C\xrightarrow{B(\_)}\mathbb{R}$ such that the equation $c_i; B(\_) = x_i$ holds. This is described by the following commutative diagram*

$$\tag{23}$$

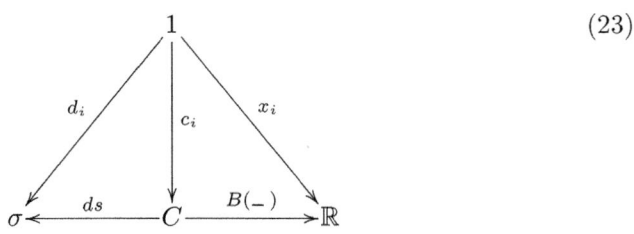

Morphism $C\xrightarrow{B(\_)}\mathbb{R}$ defines a quantitative behavior of data stream in σ^ω.

Let $\mathbb{R}^\omega = \{B(\_) \mid B(\_) : C\longrightarrow\mathbb{R}\}$. In this context, any x in $\mathbb{R}$ can also be represented as *constant* $(x, 0, 0, 0, \ldots)$ in $\mathbb{R}^\omega$, denoted by $[x]$, where $x, 0$ in $\mathbb{R}$ and its behavioral differential equation is $[x]' = [0]$. For every x in $\mathbb{R}$, a unique morphism f is defined by

$$[x]=(x,0,0,\ldots)$$

$$1 \xrightarrow{\quad x \quad} \mathbb{R} \xrightarrow{\quad f \quad} \mathbb{R}^{\omega}$$

Such that the following equation holds

$$x; f = (x, 0, 0, \ldots)$$

Sum (denoted by $\oplus$) between $B1(\_) = (x_0, x_1, \ldots)$ and $B2(\_) = (y_0, y_1, \ldots)$ in $\mathbb{R}^{\omega}$ is defined by

$$\oplus : [C \xrightarrow{B(\_)} \mathbb{R}] \times [C \xrightarrow{B(\_)} \mathbb{R}] \longrightarrow [C \xrightarrow{B(\_)} \mathbb{R}]$$

$$((B1(\_) \oplus B2(\_))(0), \ldots, (B1(\_) \oplus B2(\_))(n), \ldots) = (x_0 + y_0, \ldots, x_n + y_n, \ldots)$$

Convolution product (denoted by $\otimes$) between $B1(\_) = (x_0, x_1, \ldots)$ and $B2(\_) = (y_0, y_1, \ldots)$ in $\mathbb{R}^{\omega}$ is defined by

$$\otimes : [C \xrightarrow{B(\_)} \mathbb{R}] \times [C \xrightarrow{B(\_)} \mathbb{R}] \longrightarrow [C \xrightarrow{B(\_)} \mathbb{R}]$$

$$((B1(\_) \otimes B2(\_))(0), \ldots, (B1(\_) \otimes B2(\_))(n), \ldots) = (x_0 \times y_0, \ldots, \sum_{k=0}^{n} x_{n-k} \times y_k, \ldots).$$

From now on, we use the notations of $(\_)'$, $(\_)^{\langle n \rangle}$ and $1 \xrightarrow{0} (\_)$ for denoting $Tail(\_)$, $Tail^{\langle n \rangle}(\_)$ and $Head(\_)$ respectively. Let $X = (0, 1, 0, 0, \ldots)$ in $\mathbb{R}^{\omega}$ be *formal variable*, then it is easy to verify $X \otimes B(\_) = (B(\_))'$, $(X \otimes B(\_))' = B(\_)$ and $X \otimes B(\_) = B(\_) \otimes X$.

Let $B(d_0)$ in $\mathbb{R}^{\omega}$ be the quantitative behavior of data stream at $d_0 \in \sigma$ in DASNs. We can evaluate $B(d_0)$ in two equivalent ways as the following.

First way: Let Out be an output function defined by

$$Out : \sigma \longrightarrow \mathbb{R} \tag{24}$$

Then there is the following formula, for any $k \geqslant 0$

$$B(d_0)(k) = \sum \{x_0 \times \ldots \times x_k \times y | \exists d_0, \ldots, d_k : d_0 \xrightarrow{c_0 | x_0} \ldots \xrightarrow{c_k | x_k} d_k \stackrel{y}{\rightsquigarrow} \} \tag{25}$$

This formula computes the k-th element $B(d_0)(k)$ of $B(d_0)$ as a weighted count of the number of paths (i.e., sequences of evolutions) of length k that start in the data d_0 and end in a data with a non-zero output (i.e., $Out(d_k) = y \neq 0$ and denoted by $d_k \stackrel{y}{\rightsquigarrow}$).

Note that although the formula in (25) provides operational behavior of data stream but does not produce a compact representation for $B(d_0)$ and, hence, is not quite suitable for formal reasoning. For solving this problem, a following alternative way is useful.

Second way: Let d_1 in $\{d_{1,1}, \ldots, d_{1,n}\}$ and d_0 be data for which $next(d_0, c_0)(d_1, c_1) \neq 0$. Then the quantitative behavior $B(d_0)$ of data stream at d_0 can be completely defined by the following system of behavioral differential equations, one for each data d_0, d_1 in $\{d_{1,1}, \ldots, d_{1,n}\}$ and so on.

$$
\begin{cases}
\text{Differential equation: } (B(d_0))' = next(d_0, c_0)(d_{1,1}, c_1) \otimes B(d_{1,1}) \\
\qquad\qquad\qquad\qquad \oplus \ldots \oplus next(d_0, c_0)(d_{1,n}, c_1) \otimes B(d_{1,n}) \\
\text{Initial value:} \qquad\qquad 1 \overset{0}{\Longrightarrow} (B(d_0)) = Out(d_0)
\end{cases}
\tag{26}
$$

Theorem 6. *System of behavioral differential equations* (26) *exists a unique solution. That is, for each $B(d_0)$ there exists a unique morphism denoted by (the same notation) $B(d_0) : (\mathbb{R}^\omega)^n \longrightarrow \mathbb{R}^\omega$ satisfying* (26).

Proof. Let $TermSet$ be the set of all instances of right side in the equation (26). The uniqueness of solution stems, in fact, from there exists a unique homomorphism $h : TermSet \longrightarrow \mathbb{R}^\omega$, which assigns to each member in $TermSet$ a sequence of data in $\mathbb{R}^\omega$. Then, this sequence of data is what we define to be the effect of the morphism $B(d_0) : (\mathbb{R}^\omega)^n \longrightarrow \mathbb{R}^\omega$.   Q.E.D.

It follows from theorem 5 in section 4 that, for the quantitative behavior, representation (22) of data stream becomes $B(d_0) = 1 \overset{0}{\Longrightarrow} (B(d_0)) \oplus (X \otimes (B(d_0))')$. Thus, system of behavioral differential equations (26) is equivalent to the following system of equations.

$$
B(d_0) = Out(d_0) \oplus next(d_0, c_0)(d_{1,1}, c_1) \otimes X \otimes B(d_{1,1}) \oplus \ldots \oplus
$$
$$
next(d_0, c_0)(d_{1,n}, c_1) \otimes X \otimes B(d_{1,n})
\tag{27}
$$

Theorem 7. *For quantitative behavior of data stream, the representation* (25) *and the system of behavioral differential equations* (26) *are equivalent.*

Proof. In fact, we can represent the differential equation for $B(d_0)(k)$ in (25) as follows:

$$
\begin{aligned}
B(d_0)(k) &= B(d_0)^{\langle k \rangle}(0) \\
&= (B(d_0)')^{\langle k-1 \rangle}(0) \\
&= (next(d_0, c_0)(d_{1,1}, c_1) \times B(d_{1,1}) + \ldots + \\
&\qquad next(d_0, c_0)(d_{1,n}, c_1) \times B(d_{1,n}))^{\langle k-1 \rangle}(0) \\
&= next(d_0, c_0)(d_{1,1}, c_1) \times B(d_{1,1})^{\langle k-1 \rangle}(0) + \ldots + \\
&\qquad next(d_0, c_0)(d_{1,n}, c_1) \times B(d_{1,n})^{\langle k-1 \rangle}(0) \\
&= next(d_0, c_0)(d_{1,1}, c_1) \times B(d_{1,1})(k-1) + \ldots + \\
&\qquad next(d_0, c_0)(d_{1,n}, c_1) \times B(d_{1,n})(k-1)
\end{aligned}
$$

By induction on k, the equivalence between (25) and (26) is completely proven. Q.E.D.

6   Monoids and Streams of Data Aware Self-organizations

In this section, we construct monoids of data aware self-organizations and then streams of data aware self-organizations.

6.1   Monoids of Data Aware Self-organizations

$\mathbf{Next}_n(X, X)$, with $n \geq 0$, is a set of data aware self-organization relations $next : \sigma^n \times X \longrightarrow \sigma^n \times X$. In other words, given $n \geq 0$ then

$$\mathbf{Next}_n(X, X) = \{next : \sigma^n \times X \longrightarrow \sigma^n \times X\}$$
$$= \{next_i^{(\sigma^n \times X, \sigma^n \times X)}, \forall i \in T\}$$

Note that, in the current context, we write $next_i^{(n)}$ as a natation of $next_i^{(\sigma^n \times X, \sigma^n \times X)}$. Thus, we have

$$\mathbf{Next}_n(X, X) = \{next_i^{(n)}, \forall i \in T\}$$

This set with the composition operation "$;$" satisfies two following properties:

Relation Composition: Let f and g be members of $\mathbf{Next}_n(X, X)$, then the relation composition $(f; g) : \sigma^n \times X \longrightarrow \sigma^n \times X$ is as $g : (f : \sigma^n \times X \longrightarrow \sigma^n \times X) \longrightarrow \sigma^n \times X$. In other words, let $f = next_i^{(\sigma^n \times X, \sigma^n \times X)}$ and $g = next_j^{(\sigma^n \times X, \sigma^n \times X)}$ then

$$(next_i \; ; \; next_j)^{(\sigma^n \times X, \sigma^n \times X)} = next_j^{(next_i^{(\sigma^n \times X, \sigma^n \times X)}, \sigma^n \times X)}$$

Identity Relation: There exists an identity relation $1_{\sigma^n \times X}$ in $\mathbf{Next}_n(X, X)$ for $1_{\sigma^n \times X}; f = f; 1_{\sigma^n \times X} = f$ to be held. In other words, this can be specified by

$$next_i^{(1_{\sigma^n \times X}, \sigma^n \times X)} = next_i^{(\sigma^n \times X, 1_{\sigma^n \times X})} = next_i^{(\sigma^n \times X, \sigma^n \times X)}$$

Thus, $\mathbf{Next}_n(X, X)$ with the composition operation "$;$" is called *monoid of data aware self-organizations*. Moreover, the monoid $\mathbf{Next}_n(X, X)$ is also a monoid category including only one object to be the set $\{next : \sigma^n \times X \longrightarrow \sigma^n \times X\}$, every member of the set is a morphism, and by the composition operation the associativity and identity on the morphisms are completely satisfied.

6.2   Streams of Data Aware Self-organizations

A number of different notations are in use for denoting stream of data aware self-organizations.

$$f = (next_0^{(n)}, next_1^{(n)}, next_2^{(n)}, \dots) \tag{28}$$

is a common notation which specifies the stream f indexed by the natural numbers. We are also accustomed to

$$f = (next_i^{(n)}) \qquad \text{for all } i \text{ in } T \tag{29}$$

Informally, such a stream can be understood as a rope on which we hang up a sequence of data aware self-organizations for observation.

Note that $1 \xrightarrow{next_i^{(n)}} \mathbf{Next}_n(X, X)$ denotes for $next_i^{(n)} \in \mathbf{Next}_n(X, X)$. Hence it follows that

Definition 12. *For morphisms* $1 \xrightarrow{i} T$ *and* $1 \xrightarrow{next_i^{(n)}} \mathbf{Next}_n(X, X)$, *there exists a unique morphism* $T \xrightarrow{f} \mathbf{Next}_n(X, X)$ *such that the equation* $i; f = next_i^{(n)}$ *holds. This is described by the following commutative diagram*

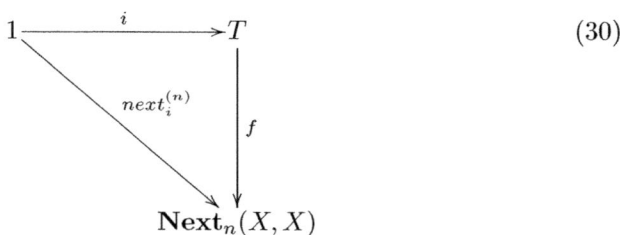

$$(30)$$

Morphism $T \xrightarrow{f} \mathbf{Next}_n(X, X)$ *defines a stream of data aware self-organizations.*

Note that morphism $T \xrightarrow{f} \mathbf{Next}_n(X, X)$ is understood as

$$\forall i[i \in T \implies \exists!\, next_i^{(n)}[next_i^{(n)} \in \mathbf{Next}_n(X, X)\ \&\ f(i) = next_i^{(n)}]]$$

In other words, $T \xrightarrow{f} \mathbf{Next}_n(X, X)$ generates stream as an infinite sequence of $f(0) = next_0^{(n)}$, $f(1) = next_1^{(n)}$, ..., $f(i) = next_i^{(n)}$, ... which is written as $\langle f(0), f(1), \dots, f(i), \dots \rangle$ or $\langle next_0^{(n)}, next_1^{(n)}, \dots, next_i^{(n)}, \dots \rangle$.

Definition 13. *Given* $T \xrightarrow{f} \mathbf{Next}_n(X, X)$ *then the set of streams, denoted by* $\mathbf{Next}_n^{\omega}(X, X)$, *is defined by*

$$\mathbf{Next}_n^{\omega}(X, X) = \{ f \mid T \xrightarrow{f} \mathbf{Next}_n(X, X) \} \tag{31}$$

The following result stems immediately from definitions 12 and 13

$$\frac{T \xrightarrow{f} \mathbf{Next}_n(X, X)}{1 \xrightarrow{f} \mathbf{Next}_n^{\omega}(X, X)} \tag{32}$$

Rule (32) means that for each morphism $T \xrightarrow{f} \mathbf{Next}_n(X, X)$, there is a morphism $1 \xrightarrow{f} \mathbf{Next}_n^{\omega}(X, X)$ generating member in $\mathbf{Next}_n^{\omega}(X, X)$. That is, morphism $T \xrightarrow{f} \mathbf{Next}_n(X, X)$ generates streams and $1 \xrightarrow{f} \mathbf{Next}_n^{\omega}(X, X)$ constructs the set of streams.

For streams we define a mechanism to generate them. This mechanism consists of an object T equipping with structural morphisms $1 \xrightarrow{0} T \xrightarrow{in\_succ} T$ with the property that for $\mathbf{Next}_n(X, X)$ and any $next_0^{(n)} \in \mathbf{Next}_n(X, X), \mathbf{Next}_n(X, X)$

$\xrightarrow{\;next\_after\;}$ $\mathbf{Next}_n(X, X)$ there exists a unique morphism $T \xrightarrow{\;f\;} \mathbf{Next}_n(X, X)$ such that the following diagram commutes

$$
\begin{array}{ccc}
1 \xrightarrow{\;\;0\;\;} T \xrightarrow{\;\;in\_succ\;\;} T & & (33)\\
\end{array}
$$

Definition 14. *We define a stream construction morphism, denoted by* $\ddagger$, *such that*

$$\mathbf{Next}_n(X, X) \times [T \xrightarrow{\;f\;} \mathbf{Next}_n(X, X)] \xrightarrow{\;\ddagger\;} [T \xrightarrow{\;f\;} \mathbf{Next}_n(X, X)]$$

This means that

$$\ddagger(A \times B \xrightarrow{\;f \times g\;} C \times D) = A \ddagger B \xrightarrow{\;f \ddagger g\;} C \ddagger D$$

It follows that any stream $T \xrightarrow{\;f\;} \mathbf{Next}_n(X, X)$ can be represented in a format including two parts of *head* and *tail* that are connected by "$\ddagger$" such that

$$T \xrightarrow{\;f\;} \mathbf{Next}_n(X, X) \equiv$$

where $1 \xrightarrow{\;0\;} T \xrightarrow{\;f\;} \mathbf{Next}_n(X, X) = f(0)$ and $1 \xrightarrow{\;i \geqslant 0\;} T \xrightarrow{\;f\;} \mathbf{Next}_n(X, X) = \langle f(1), f(2), \ldots \rangle$ are called head and tail, respectively.

Definition 15. *We define a head construction morphism, denoted by* $1 \overset{0}{\Longrightarrow} (\_)$, *such that*

$$[T \xrightarrow{\;f\;} \mathbf{Next}_n(X, X)] \xrightarrow{\;1 \overset{0}{\Rightarrow}(\_)\;} \mathbf{Next}_n(X, X)$$

This states that $\forall(a \ddagger s)[(a \ddagger s) \in [T \xrightarrow{\;f\;} \mathbf{Next}_n(X, X)] \implies \exists! next_0^{(n)}[next_0^{(n)} \in \mathbf{Next}_n(X, X) \;\&\; 1 \overset{0}{\Longrightarrow}(a \ddagger s) = a = next_0^{(n)}]]$.

It follows that $1 \overset{0}{\Longrightarrow}(T \xrightarrow{\;f\;} \mathbf{Next}_n(X, X)) \equiv 1 \xrightarrow{\;0\;} T \xrightarrow{\;f\;} \mathbf{Next}_n(X, X)$.

Definition 16. *We define a tail construction morphism, denoted by* $(\_)'$*, such that*

$$[T \xrightarrow{f} \mathbf{Next}_n(X, X)] \xrightarrow{\quad (\_)' \quad} [T \xrightarrow{f} \mathbf{Next}_n(X, X)]$$

This means that $\forall (a \ddagger s)[(a \ddagger s) \in [T \xrightarrow{f} \mathbf{Next}_n(X, X)] \implies \exists! \langle next_1^{(n)}, next_2^{(n)}, \dots \rangle [\langle next_1^{(n)}, next_2^{(n)}, \dots \rangle \in [T \xrightarrow{f} \mathbf{Next}_n(X, X)] \& (a \ddagger s)' = s = \langle next_1^{(n)}, next_2^{(n)}, \dots \rangle]]$.

As a convention, $(\_)^{\langle n \rangle}$ denotes applying recursively the $(\_)'$ n times. Thus, specifically, $(\_)^{\langle 2 \rangle}, (\_)^{\langle 1 \rangle}$ and $(\_)^{\langle 0 \rangle}$ stand for $((\_)')', (\_)'$ and $(\_)$, respectively.

It follows that the first member of stream $T \xrightarrow{f} \mathbf{Next}_n(X, X)$ is given by

$$1 \xRightarrow{0} ((T \xrightarrow{f} \mathbf{Next}_n(X, X))') \equiv 1 \xrightarrow{1} T \xrightarrow{f} \mathbf{Next}_n(X, X)$$

and, in general, for every $k \in T$ the k-th member of stream $T \xrightarrow{f} \mathbf{Next}_n(X, X)$ is provided by

$$1 \xRightarrow{0} ((T \xrightarrow{f} \mathbf{Next}_n(X, X))^{\langle k \rangle}) \equiv 1 \xrightarrow{k} T \xrightarrow{f} \mathbf{Next}_n(X, X) \qquad (34)$$

Stream to be an infinite sequence of all $next_i^{(n)}$ is viewed and treated as single mathematical entity, so the derivative of stream $T \xrightarrow{f} \mathbf{Next}_n(X, X)$ is given by $(T \xrightarrow{f} \mathbf{Next}_n(X, X))'$.

Now using this notation for stream derivative, we can specify stream $T \xrightarrow{f} \mathbf{Next}_n(X, X)$ as

- Initial value: $1 \xrightarrow{0} T \xrightarrow{f} \mathbf{Next}_n(X, X)$ and
- Differential equation:

$$((T \xrightarrow{f} \mathbf{Next}_n(X, X))^{\langle n \rangle})' = (T \xrightarrow{f} \mathbf{Next}_n(X, X))^{\langle n+1 \rangle}$$

The initial value of $T \xrightarrow{f} \mathbf{Next}_n(X, X)$ is defined as its first element $1 \xrightarrow{0} T \xrightarrow{f} \mathbf{Next}_n(X, X)$, and the stream derivative, denoted by $(T \xrightarrow{f} \mathbf{Next}_n(X, X))'$, is defined by $((T \xrightarrow{f} \mathbf{Next}_n(X, X))^{\langle n \rangle})' = (T \xrightarrow{f} \mathbf{Next}_n(X, X))^{\langle n+1 \rangle}$, for any integer $n \in T$. In other words, the initial value and derivative equal the head and tail of $T \xrightarrow{f} \mathbf{Next}_n(X, X)$, respectively. The behavior of a stream $T \xrightarrow{f} \mathbf{Next}_n(X, X)$ consists of two aspects: it allows for the observation of its initial value $1 \xrightarrow{0} T \xrightarrow{f} \mathbf{Next}_n(X, X)$; and it can make an evolution to the new stream $(T \xrightarrow{f} \mathbf{Next}_n(X,X))'$, consisting of the original stream from which the first element has been removed. The initial value of $(T \xrightarrow{f} \mathbf{Next}_n(X, X))'$, which is

$$1 \xRightarrow{0} ((T \xrightarrow{f} \mathbf{Next}_n(X,X))') = 1 \xrightarrow{1} T \xrightarrow{f} \mathbf{Next}_n(X,X)$$

can in its turn be observed, but note that we had to move from $T \xrightarrow{f} \mathbf{Next}_n(X, X)$ to $(T \xrightarrow{f} \mathbf{Next}_n(X,X))'$ first in order to do so. Now a behavioral differential equation defines a stream by specifying its initial value together with a description of its derivative, which tells us how to continue.

Every member $next_i^{(n)}$ in $\mathbf{Next}_n(X,X)$ can be considered as a stream in the following manner. For every $next_i^{(n)}$ in $\mathbf{Next}_n(X,X)$, a unique stream is defined by morphism f:

$$1 \xrightarrow{next_i^{(n)}} \mathbf{Next}_n(X,X) \xrightarrow{f} \mathbf{Next}_n^{\omega}(X,X)$$

with the overarching $\langle next_i^{(n)}, 1_{\sigma^n \times X}, 1_{\sigma^n \times X}, \ldots \rangle$

Such that the following equation holds

$$next_i^{(n)}; f = \langle next_i^{(n)}, 1_{\sigma^n \times X}, 1_{\sigma^n \times X}, \ldots \rangle$$

with $\langle next_i^{(n)}, 1_{\sigma^n \times X}, 1_{\sigma^n \times X}, \ldots \rangle$ in $\mathbf{Next}_n^{\omega}(X,X)$.

Definition 17. *For any* $T \xrightarrow{f} \mathbf{Next}_n(X,X)$ *and* $T \xrightarrow{g} \mathbf{Next}_n(X,X)$, *we define*

$$f = g \text{ iff } 1 \xrightarrow{n} T \xrightarrow{f} \mathbf{Next}_n(X,X) = 1 \xrightarrow{n} T \xrightarrow{g} \mathbf{Next}_n(X,X) \quad (35)$$

with every $n \in T$.

Definition 18. *Bisimulation on* $\mathbf{Next}_n^{\omega}(X,X)$ *is a relation, denoted by* $\sim$, *between streams* $T \xrightarrow{f} \mathbf{Next}_n(X,X)$ *and* $T \xrightarrow{g} \mathbf{Next}_n(X,X)$ *such that*

$$f \sim g$$
$$1 \xRightarrow{0} (f) = 1 \xRightarrow{0} (g) \text{ and } (f)' \sim (g)' \quad (36)$$

Two streams of data aware self-organizations are bisimular if, regarding their behaviors, each of the streams "simulates" the other and vice-versa. In other words, each of the streams cannot be distinguished from the other by the observation. We obtain

Corollary 1. *Let f, g and h be in $\mathbf{Next}_n^\omega(X, X)$. If $f \sim g$ and $g \sim h$ then $(f \sim g) \circ (g \sim h) = f \sim h$, where the symbol $\circ$ denotes a relational composition. For more descriptive notation, we can write this in the form*

$$\frac{f \sim g, g \sim h}{(f \sim g) \circ (g \sim h) = f \sim h} \tag{37}$$

and conversely, if $f \sim h$ then there exists g such that $f \sim g$ and $g \sim h$. This can be written as

$$\frac{f \sim h}{\exists g : f \sim g \quad \text{and} \quad g \sim h} \tag{38}$$

Proof. Proving (37) originates as the result of the truth that the relational composition between two bisimulations $L_1 \subseteq f \times g$ and $L_2 \subseteq g \times h$ is a bisimulation obtained by $L_1 \circ L_2 = \{\langle x, y\rangle \mid x\, L_1\, z \text{ and } z\, L_2\, y \text{ for some } z \in g\}$, where $x \in f$, $z \in g$ and $y \in h$.

Proving (38) comes from the fact that there are always $g = f$ or $g = h$ as simply as they can. Hence, (38) is always true in general.   Q.E.D.

Corollary 2. *Let $f_i, \forall i \in \mathbb{N}$, be in $\mathbf{Next}_n^\omega(X, X)$ and $\bigcup\limits_{i \in \mathbb{N}}$ be union of a family of sets. Then*

$$\frac{f \sim f_i \quad \text{with } i \in \mathbb{N}}{\bigcup\limits_{i \in \mathbb{N}} (f \sim f_i) = f \sim \bigcup\limits_{i \in \mathbb{N}} f_i} \tag{39}$$

and conversely,

$$\frac{f \sim \bigcup\limits_{i \in \mathbb{N}} f_i}{\exists i \in \mathbb{N} : f \sim f_i} \tag{40}$$

Proof. Proving (39) stems straightforwardly from the fact that f bisimulates f_i (i.e., $f \sim f_i$), then f bisimulates each stream in $\bigcup\limits_{i \in \mathbb{N}} f_i$.

Conversely, proving (40) develops as the result of the fact that for each $\langle x, y\rangle \in \bigcup\limits_{i \in \mathbb{N}} (f \times f_i)$, there exists $i \in \mathbb{N}$ such that $\langle x, y\rangle \in f \times f_i$. In other words, it is formally denoted by $\bigcup\limits_{i \in \mathbb{N}} (f \times f_i) = \{\langle x, y\rangle \mid \exists i \in \mathbb{N} : x \in f \quad \text{and} \quad y \in f_i\}$, where $x \in f$ and $y \in f_i$.   Q.E.D.

The union of all bisimulations between f and f_i (i.e., $\bigcup\limits_{i \in \mathbb{N}} (f \sim f_i)$) is the greatest bisimulation. The greatest bisimulation is called the *bisimulation equivalence* or *bisimilarity* [17,10] (again denoted by the notation $\sim$).

Corollary 3. *The bisimilarity $\sim$ on $\bigcup\limits_{i \in \mathbb{N}} (f \sim f_i)$ is an equivalence relation.*

Proof. In fact, a bisimilarity $\sim$ on $\bigcup\limits_{i \in \mathbb{N}} (f \sim f_i)$ is a binary relation $\sim$ on $\bigcup\limits_{i \in \mathbb{N}} (f \sim f_i)$, which is reflexive, symmetric and transitive. In other words, the following properties hold for $\sim$.

– Reflexivity:

$$\frac{\forall (a \sim b) \in \bigcup_{i \in \mathbb{N}} (f \sim f_i)}{(a \sim b) \sim (a \sim b)} \tag{41}$$

– Symmetry:

$$\begin{array}{c} \forall (a \sim b), (c \sim d) \in \bigcup_{i \in \mathbb{N}} (f \sim f_i), \\ (a \sim b) \sim (c \sim d) \\ \hline (c \sim d) \sim (a \sim b) \end{array} \tag{42}$$

– Transitivity:

$$\begin{array}{c} \forall (a \sim b), (c \sim d), (e \sim f) \in \bigcup_{i \in \mathbb{N}} (f \sim f_i), \\ ((a \sim b) \sim (c \sim d)) \bigwedge ((c \sim d) \sim (e \sim f)) \\ \hline (a \sim b) \sim (e \sim f) \end{array} \tag{43}$$

to be an equivalence relation on $\bigcup_{i \in \mathbb{N}} (f \sim f_i)$.   Q.E.D.

For some constraint α, if $g \sim h$ then two streams g and h have the following relation.

$$\frac{g \models \alpha}{h \models \alpha} \tag{44}$$

That is, if stream g satisfies constraint α then this constraint is still preserved on stream h. Thus it is read as $g \sim h$ in the constraint of α (and denoted by $g \sim_\alpha h$).

For validating whether $f = g$, a powerful method is so-called *proof principle of coinduction* [18] that states as follows:

Theorem 8

$$\frac{f \sim g}{f = g} \tag{45}$$

Proof. Hence in order to prove the equality of two streams f and g, it is sufficient to establish the existence of a bisimulation relation $f \sim g$.

In fact, for two streams f and g and a bisimulation $f \sim g$. We have

By induction on $n \geqslant 0$ and bisimulation $\sim$
$\vdash f^{\langle n \rangle} = g^{\langle n \rangle}$
By bisimulation $\sim$
$\vdash 1 \overset{0}{\Longrightarrow} (f^{\langle n \rangle}) = 1 \overset{0}{\Longrightarrow} (g^{\langle n \rangle})$
By identity in (34)
$\vdash 1 \overset{n}{\longrightarrow} f = 1 \overset{n}{\longrightarrow} g$ with every $n \geqslant 0$
It follows that, by (35), $f = g$.   Q.E.D.

As a consequence, using coinduction we can establish the validity of the equivalence between streams $f = T \overset{f}{\longrightarrow} \mathbf{Next}_n(X, X)$ and $g = T \overset{g}{\longrightarrow} \mathbf{Next}_n(X, X)$ in $\mathbf{Next}_n^\omega(X, X)$.

Theorem 9. *For every* $f \in \mathbf{Next}_n^\omega(X, X)$,

$$f = 1\overset{0}{\Longrightarrow}(f) \ddagger (f)' \tag{46}$$

Proof. This stems from the coinductive proof principle in (45). In fact, it is easy to check the following bisimulation $f \sim 1\overset{0}{\Longrightarrow}(f) \ddagger (f)'$. It follows that $f = 1\overset{0}{\Longrightarrow}(f) \ddagger (f)'$.   Q.E.D.

In (46), operation $\ddagger$ as a kind of stream integration, the theorem states that stream derivation and stream integration are inverse operations. It gives a way to obtain f from $(f)'$ and the initial value $1\overset{0}{\Longrightarrow}(f)$. As a result, the theorem allows us to reach solution of differential equations in an algebraic manner.

6.3   Polymorphic Stream Construction

We see that morphism $\mathbf{Next}_n(X, X) \ddagger \mathbf{Next}_n^\omega(X, X) \overset{f}{\longrightarrow} \mathbf{Next}_n^\omega(X, X)$ turns out to be as a *stream generating algebra* with a structure of functor to be "$\mathbf{Next}_n(X, X)\ddagger\_$" that maps streams in $\mathbf{Next}_n^\omega(X, X)$ to others in $\mathbf{Next}_n^\omega(X, X)$. We now abstract away from the initial choice of $\mathbf{Next}_n^\omega(X, X)$. Given the functor "$\mathbf{Next}_n(X, X) \ddagger \_$", we construct a polymorphic stream to the family of streams induced by the above-mentioned functor.

In fact, let $\mathbf{Q}$ be an arbitrary category and given a morphism $\mathbf{Next}_n(X, X) \ddagger \mathbf{Q} \overset{g}{\longrightarrow} \mathbf{Q}$, there exists a unique $\mathbf{Next}_n(X, X)\ddagger$-homomorphism ψ which maps every object in $\mathbf{Q}$ to a stream in $\mathbf{Next}_n^\omega(X, X)$ such that the following equation holds

$$1_{\mathbf{Next}_n(X,X)} \ddagger \psi; f = g; \psi \tag{47}$$

This is equivalent to saying that the following diagram commutes.

$$
\begin{array}{ccc}
\mathbf{Next}_n(X, X) \ddagger \mathbf{Q} & \overset{g}{\longrightarrow} & \mathbf{Q} \\
{\scriptstyle 1_{\mathbf{Next}_n(X,X)}\ddagger\psi}\big\downarrow & & \big\downarrow{\scriptstyle \psi} \\
\mathbf{Next}_n(X, X) \ddagger \mathbf{Next}_n^\omega(X, X) & \overset{f}{\longrightarrow} & \mathbf{Next}_n^\omega(X, X)
\end{array} \tag{48}
$$

Now we justify this statement. To see what $\mathbf{Q}$ is, we have to determine the process that starts from an object q in $\mathbf{Q}$. That is, we have to analyze step-by-step that we can reach q from which objects and by which data aware self-organizations.

$$q$$

$$\mathbf{Next}_n(X, X) \ddagger q^{\langle 1 \rangle} \overset{g}{=} q$$

$$[\mathbf{Next}_n(X,X)\ddagger]^2 q^{\langle 2 \rangle} \stackrel{g}{=} \mathbf{Next}_n(X,X) \ddagger q^{\langle 1 \rangle} \stackrel{g}{=} q$$

$$\cdots$$

$$[\mathbf{Next}_n(X,X)\ddagger]^i q^{\langle i \rangle} \stackrel{g}{=} [\mathbf{Next}_n(X,X)\ddagger]^{i-1} q^{\langle i-1 \rangle} \stackrel{g}{=} \ldots \stackrel{g}{=} q$$

$$\cdots$$

where

- Notation $\stackrel{g}{=}$ stands for an equal relation up to g,
- $q, q^{\langle 1 \rangle}, \ldots, q^{\langle i \rangle}, \ldots$ are objects which occur in the process

$$\cdots q^{\langle i \rangle} \cdots \xrightarrow{\mathbf{Next}_n(X,X)\ddagger} q^{\langle 2 \rangle} \xrightarrow{\mathbf{Next}_n(X,X)\ddagger} q^{\langle 1 \rangle} \xrightarrow{\mathbf{Next}_n(X,X)\ddagger} q,$$

- $[\mathbf{Next}_n(X,X)\ddagger]^2$ means $\mathbf{Next}_n(X,X) \ddagger \mathbf{Next}_n(X,X)\ddagger$ and
- $[\mathbf{Next}_n(X,X)\ddagger]^i$ means $[\mathbf{Next}_n(X,X)\ddagger]^{i-1}\mathbf{Next}_n(X,X)\ddagger$

The unfolding of the process $\mathbf{Next}_n(X,X) \ddagger \mathbf{Q} \xrightarrow{g} \mathbf{Q}$ gives the following sequence:

$$\cdots \xrightarrow{[\mathbf{Next}_n(X,X)\ddagger]^i g} [\mathbf{Next}_n(X,X)\ddagger]^i \mathbf{Q} \xrightarrow{[\mathbf{Next}_n(X,X)\ddagger]^{i-1} g} \cdots [\mathbf{Next}_n(X,X)\ddagger]^2 \mathbf{Q}$$

$$\xrightarrow{\mathbf{Next}_n(X,X)\ddagger g} \mathbf{Next}_n(X,X) \ddagger \mathbf{Q} \xrightarrow{g} \mathbf{Q}$$

In this sequence, for any object q in $\mathbf{Q}$ then $\mathbf{Next}_n(X,X) \ddagger \mathbf{Q} \xrightarrow{g} \mathbf{Q}$ describes that q can be reached from which objects in one step and by which data aware self-organizations.

$[\mathbf{Next}_n(X,X)\ddagger]^i \mathbf{Q} \xrightarrow{[\mathbf{Next}_n(X,X)\ddagger]^{i-1} g} [\mathbf{Next}_n(X,X)\ddagger]^{i-1} \mathbf{Q}$ describes how arbitrary sequence of data aware self-organizations of length i can be continued according to g in the next step. Starting from an object q we obtain in such a way an infinite sequence $q \stackrel{g}{=} \mathbf{Next}_n(X,X) \ddagger q^{\langle 1 \rangle} \stackrel{g}{=} [\mathbf{Next}_n(X,X)\ddagger]^2 q^{\langle 2 \rangle} \stackrel{g}{=} [\mathbf{Next}_n(X,X)\ddagger]^3 q^{\langle 3 \rangle} \stackrel{g}{=} \ldots \stackrel{g}{=} [\mathbf{Next}_n(X,X)\ddagger]^i q^{\langle i \rangle} \stackrel{g}{=} \ldots$.

We define $\mathbf{Next}_n(X,X)\ddagger$-homomorphism ψ that assigns every object q in $\mathbf{Q}$ to a stream $\langle 1 \stackrel{0}{\Longrightarrow} (q), 1 \stackrel{0}{\Longrightarrow} (q^{\langle 1 \rangle}), 1 \stackrel{0}{\Longrightarrow} (q^{\langle 2 \rangle}), 1 \stackrel{0}{\Longrightarrow} (q^{\langle 3 \rangle}), \ldots, 1 \stackrel{0}{\Longrightarrow} (q^{\langle i \rangle}), \ldots \rangle$ in $\mathbf{Next}_n^\omega(X,X)$. That is,

$$\mathbf{Q} \xrightarrow{\psi} \mathbf{Next}_n^\omega(X,X)$$

$$q \xmapsto{\psi} \langle 1 \stackrel{0}{\Longrightarrow} (q), 1 \stackrel{0}{\Longrightarrow} (q^{\langle 1 \rangle}), 1 \stackrel{0}{\Longrightarrow} (q^{\langle 2 \rangle}), \ldots, 1 \stackrel{0}{\Longrightarrow} (q^{\langle i \rangle}), \ldots \rangle$$

or

$$\langle 1 \stackrel{0}{\Longrightarrow} (q), 1 \stackrel{0}{\Longrightarrow} (q^{\langle 1 \rangle}), 1 \stackrel{0}{\Longrightarrow} (q^{\langle 2 \rangle}), \ldots, 1 \stackrel{0}{\Longrightarrow} (q^{\langle i \rangle}), \ldots \rangle$$

$$1 \xrightarrow{q} \mathbf{Q} \xrightarrow{\psi} \mathbf{Next}_n^\omega(X,X)$$

or

$$\psi(q) = \langle 1 \stackrel{0}{\Longrightarrow} (q), 1 \stackrel{0}{\Longrightarrow} (q^{\langle 1 \rangle}), 1 \stackrel{0}{\Longrightarrow} (q^{\langle 2 \rangle}), \ldots, 1 \stackrel{0}{\Longrightarrow} (q^{\langle i \rangle}), \ldots \rangle$$

It follows that by above-mentioned process to construct ψ it is easy to check ψ satisfying the property (47) of homomorphism. The following is one step of checking $(q^{\langle 1 \rangle} \xrightarrow{\mathbf{Next}_n(X,X)\ddagger} q)$.

$$g(\mathbf{Next}_n(X,X)\ddagger q^{\langle 1 \rangle}) = q$$

$$\psi(q) = \langle 1 \xRightarrow{0} (q), 1 \xRightarrow{0} (q^{\langle 1 \rangle}), 1 \xRightarrow{0} (q^{\langle 2 \rangle}), \ldots, 1 \xRightarrow{0} (q^{\langle i \rangle}), \ldots \rangle$$

$$1_{\mathbf{Next}_n(X,X)} \ddagger \psi(q^{\langle 1 \rangle}) = \langle 1 \xRightarrow{0} (q^{\langle 1 \rangle}), 1 \xRightarrow{0} (q^{\langle 2 \rangle}), \ldots, 1 \xRightarrow{0} (q^{\langle i \rangle}), \ldots \rangle$$

$$f(\mathbf{Next}_n(X,X) \ddagger \langle 1 \xRightarrow{0} (q^{\langle 1 \rangle}), 1 \xRightarrow{0} (q^{\langle 2 \rangle}), \ldots \rangle) = \langle 1 \xRightarrow{0} (q), 1 \xRightarrow{0} (q^{\langle 1 \rangle}), \ldots \rangle$$

Hence, $1_{\mathbf{Next}_n(X,X)} \ddagger \psi(q^{\langle 1 \rangle}) \; ; f(\mathbf{Next}_n(X,X)\ddagger \langle 1 \xRightarrow{0} (q^{\langle 1 \rangle}), 1 \xRightarrow{0} (q^{\langle 2 \rangle}), \ldots \rangle) = g(\mathbf{Next}_n(X,X) \ddagger q^{\langle 1 \rangle}) \; ; \psi(q) = \langle 1 \xRightarrow{0} (q), 1 \xRightarrow{0} (q^{\langle 1 \rangle}), 1 \xRightarrow{0} (q^{\langle 2 \rangle}), \ldots, 1 \xRightarrow{0} (q^{\langle i \rangle}), \ldots \rangle$. In other words, the following diagram commutes

$$
\begin{array}{ccc}
\mathbf{Next}_n(X,X)\ddagger q^{\langle 1 \rangle} & \xrightarrow{\;\;g\;\;} & q \\
\Big\downarrow{\scriptstyle 1_{\mathbf{Next}_n(X,X)}\ddagger\psi} & & \Big\downarrow{\scriptstyle \psi} \\
\mathbf{Next}_n(X,X)\ddagger \langle 1 \xRightarrow{0} (q^{\langle 1 \rangle}), 1 \xRightarrow{0} (q^{\langle 2 \rangle}), \ldots \rangle & \xrightarrow{\;\;f\;\;} & \langle 1 \xRightarrow{0} (q), 1 \xRightarrow{0} (q^{\langle 1 \rangle}), \ldots \rangle
\end{array}
$$

Checking the other steps is completely similar.

The uniqueness of ψ to be a unique $\mathbf{Next}_n(X,X)\ddagger$-homomorphism stems straightforwardly from the coinductive proof principle in (45). Suppose that there are two homomorphisms $\psi(q)$ and $\alpha(q)$.

Let $\sim = \{\langle \psi(q), \alpha(q) \rangle \mid q \text{ is an object in } \mathbf{Q}\}$. We prove that $\sim$ is a bisimulation on $\mathbf{Next}_n^\omega(X,X)$. Consider $\langle \psi(q), \alpha(q) \rangle \in \sim$, then

$$\psi(q) \text{ and } \alpha(q) \text{ are homomorphims } \vdash \begin{pmatrix} 1 \xRightarrow{0} (\psi(q)) \\ = \\ 1 \xRightarrow{0} (\alpha(q)) \end{pmatrix}$$

$$\psi(q) \text{ and } \alpha(q) \text{ are homomorphims } \vdash \begin{pmatrix} \psi^{\langle 1 \rangle}(q) = \psi(q^{\langle 1 \rangle}) \\ \text{and} \\ \alpha^{\langle 1 \rangle}(q) = \alpha(q^{\langle 1 \rangle}) \end{pmatrix}$$

$$\begin{pmatrix} \psi^{\langle 1 \rangle}(q) = \psi(q^{\langle 1 \rangle}) \\ \text{and} \\ \alpha^{\langle 1 \rangle}(q) = \alpha(q^{\langle 1 \rangle}) \\ \text{and} \\ \langle \psi(q^{\langle 1 \rangle}), \alpha(q^{\langle 1 \rangle}) \rangle \in \sim \end{pmatrix} \vdash \langle \psi^{\langle 1 \rangle}(q), \alpha^{\langle 1 \rangle}(q) \rangle \in \sim$$

$$\begin{pmatrix} \langle \psi^{\langle 1 \rangle}(q), \alpha^{\langle 1 \rangle}(q) \rangle \in \sim \\ \text{and} \\ 1 \xRightarrow{0} (\psi(q)) = 1 \xRightarrow{0} (\alpha(q)) \end{pmatrix} \vdash \sim \text{ is a bisimulation}$$

Note that, for any object q in $\mathbf{Q}$, if $\langle\psi(q),\alpha(q)\rangle \in\sim$ we can write $\psi(q) \sim \alpha(q)$. Thus, by the coinductive proof principle in (45) then $\psi = \alpha$.

7   Constructing a Category of Monoids $\mathbf{Next}_i(X, X)$

By the monoids $\mathbf{Next}_i(X, X)$ of data aware self-organizations, for all $i \in T$, we can construct $\mathbf{Stream}_\sigma(\mathbf{Next})(X, X)$ to be a monoids category. In fact, category $\mathbf{Stream}_\sigma(\mathbf{Next})(X, X)$ is constructed as follows:

$Obj(\mathbf{Stream}_\sigma(\mathbf{Next})(X, X))$ is a set of monoids $\mathbf{Next}_i(X, X)$ for all $i \in T$. In other words,

$$Obj(\mathbf{Stream}_\sigma(\mathbf{Next})(X, X)) = \{\mathbf{Next}_i(X, X), \forall i \in T\}$$

Associated with each object $\mathbf{Next}_i(X, X)$ in $Obj(\mathbf{Stream}_\sigma(\mathbf{Next})(X, X))$, we define a morphism $\mathbf{Next}_i(X, X) \xrightarrow{1_{\mathbf{Next}_i(X,X)}} \mathbf{Next}_i(X, X)$, the identity morphism on $\mathbf{Next}_i(X, X)$ such that

$$\mathbf{Next}_i(X, X) \xrightarrow{1_{\mathbf{Next}_i(X,X)} = 1_{\sigma^i \times X}} \mathbf{Next}_i(X, X)$$

or

$$\{next : \sigma^i \times X \longrightarrow \sigma^i \times X\} \xrightarrow{1_{\mathbf{Next}_i(X,X)} = 1_{\sigma^i \times X}} \{next : \sigma^i \times X \longrightarrow \sigma^i \times X\}$$

and to each pair of morphisms $\mathbf{Next}_i(X, X) \xrightarrow{f} \mathbf{Next}_j(X, X)$ and $\mathbf{Next}_j(X, X)$

$\xrightarrow{g} \mathbf{Next}_k(X, X)$ such that

$$\mathbf{Next}_i(X, X) \xrightarrow{f = \sigma^{j-i} \times 1_{\sigma^i \times X}} \mathbf{Next}_j(X, X)$$

and

$$\mathbf{Next}_j(X, X) \xrightarrow{g = \sigma^{k-j} \times 1_{\sigma^j \times X}} \mathbf{Next}_k(X, X)$$

there is an associated morphism $\mathbf{Next}_i(X, X) \xrightarrow{f;g} \mathbf{Next}_k(X, X)$, the composition of f with g such that

$$\mathbf{Next}_i(X, X) \xrightarrow{f;g = \sigma^{k-i} \times 1_{\sigma^i \times X}} \mathbf{Next}_k(X, X)$$

For all objects in $Obj(\mathbf{Stream}_\sigma(\mathbf{Next})(X, X))$ and the morphisms

$$\mathbf{Next}_i(X, X) \xrightarrow{f = \sigma^{j-i} \times 1_{\sigma^i \times X}} \mathbf{Next}_j(X, X)$$

$$\mathbf{Next}_j(X,X) \xrightarrow{\quad g = \sigma^{k-j} \times 1_{\sigma^j \times X} \quad} \mathbf{Next}_k(X,X)$$

and

$$\mathbf{Next}_k(X,X) \xrightarrow{\quad h = \sigma^{m-k} \times 1_{\sigma^k \times X} \quad} \mathbf{Next}_m(X,X)$$

in $Arc(\mathbf{Stream}_\sigma(\mathbf{Next})(X,X))$, the following equations hold:

Associativity: $\quad (f;g);h = f;(g;h) = \sigma^{m-i} \times 1_{\sigma^i \times X}$

Identity: $\quad 1_{\mathbf{Next}_i(X,X)};f = f = f;1_{\mathbf{Next}_j(X,X)}$

(i.e., $1_{\sigma^i \times X};\sigma^{j-i} \times 1_{\sigma^i \times X} = \sigma^{j-i} \times 1_{\sigma^i \times X} = \sigma^{j-i} \times 1_{\sigma^i \times X};1_{\sigma^j \times X}$)

As a result, the above-mentioned monoid morphisms can be diagrammatically drawn such as

$$\mathbf{Next}_i(X,X) \xrightarrow{\quad \sigma^{\pm k} \times 1_{\sigma^i \times X} \quad} \mathbf{Next}_{i \pm k}(X,X) \tag{49}$$

or

$$\{next : \sigma^i \times X \longrightarrow \sigma^i \times X\} \xrightarrow{\quad \sigma^{\pm k} \times 1_{\sigma^i \times X} \quad} \{next : \sigma^{i \pm k} \times X \longrightarrow \sigma^{i \pm k} \times X\}$$

A consequence coming from constructing $\mathbf{Stream}_\sigma(\mathbf{Next})(X,X))$ is stated by

Corollary 4. *All monoid morphisms of* $\mathbf{Stream}_\sigma(\mathbf{Next})(X,X)$ *are monoid isomorphisms*

Proof. In fact, this result immediately stems from (49). For every pair of monoid morphisms $\mathbf{Next}_i(X,X)$ and $\mathbf{Next}_j(X,X)$ in $Arc(\mathbf{Stream}_\sigma(\mathbf{Next})(X,X))$, we have

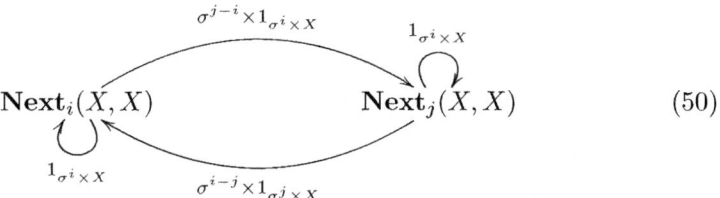

$$\tag{50}$$

These morphisms are exactly isomorphic. $\hfill$ Q.E.D.

Definition 19. *We define the monoid isomorphisms between the pair of monoids in* $\mathbf{Stream}_\sigma$ *(*$\mathbf{Next}$*)(X,X) as isomorphisms of* DASNs.

From the above-mentioned justification of $\mathbf{Stream}_\sigma(\mathbf{Next})(X,X))$, it is possible to derive $\mathbf{Next}_i(X,X)$ for every $i \in T$. Derivation of every $\mathbf{Next}_i(X,X)$ is simplified by the following inference rules of categorical programming:

$$\tag{51}$$

$$1 \xrightarrow{\quad \mathbf{Next}(X,X) \quad} Obj(\mathbf{Stream}_\sigma(\mathbf{Next})(X,X))$$

This means that it is always able to select a $\mathbf{Next}(X, X)$ from $Obj(\mathbf{Stream}_\sigma$ $(\mathbf{Next})(X, X))$ we have constructed. Note that, $\mathbf{Next}(X, X)$ $=$ $\{next : \sigma^0 \times X \longrightarrow \sigma^0 \times X\} = \{next : X \longrightarrow X\}$.

$$\left(\begin{array}{c} 1 \xrightarrow{\mathbf{Next}(X,X)} Obj(\mathbf{Stream}_\sigma(\mathbf{Next})(X, X)) \\ \text{and} \\ \mathbf{Next}(X, X) \xrightarrow{\sigma^i \times 1_X} \mathbf{Next}_i(X, X) \\ 1 \xrightarrow{\mathbf{Next}_i(X,X)} Obj(\mathbf{Stream}_\sigma(\mathbf{Next})(X, X)) \end{array} \right) \tag{52}$$

It means that given $\mathbf{Next}(X, X)$ we can compute $\mathbf{Next}_i(X, X)$ for every i. Note that $\mathbf{Next}(X, X) \xrightarrow{1_X} \mathbf{Next}(X, X)$.

$$\left(\begin{array}{c} 1 \xrightarrow{\mathbf{Next}_i(X,X)} Obj(\mathbf{Stream}_\sigma(\mathbf{Next})(X, X)) \\ \text{and} \\ \mathbf{Next}_i(X, X) \xrightarrow{\sigma^{j-i} \times 1_{\sigma^i \times X}} \mathbf{Next}_j(X, X) \\ 1 \xrightarrow{\mathbf{Next}_j(X,X)} Obj(\mathbf{Stream}_\sigma(\mathbf{Next})(X, X)) \end{array} \right) \tag{53}$$

This rule states that given $\mathbf{Next}_i(X, X)$ we can compute $\mathbf{Next}_j(X, X)$ for every $j \neq i$.

From the $\mathbf{Stream}_\sigma(\mathbf{Next})(X, X)$ construction, we see that every $\mathbf{Next}_i(X, X)$ can be formed in the unifying way and, moreover, it is also easy to see that we can write code in virtually any implementation language for performing the construction procedure based on these three rules. The major point is that we have gained the substantial aspects of the construction procedure without any excessive inclination towards a specific implementation detail; it is at a high abstract level. This is quite helpful when we want to prove properties of the construction. In fact, we can prove

Theorem 10. *Every object* $\mathbf{Next}_i(X, X)$ *can be constructed by any other object in* $\mathbf{Stream}_\sigma$ $(\mathbf{Next})(X, X)$

Proof. Applying rule in (52) and/or rule in (53) to construct every object $\mathbf{Next}_i(X, X)$ from another object in $\mathbf{Stream}_\sigma(\mathbf{Next})(X, X)$ Q.E.D.

This is certainly a property we expect of any construction procedure.

Theorem 11. $\mathbf{Stream}_\sigma(\mathbf{Next})(X, X)$ *is a complete graph*

Proof. In fact, this is a consequence stemming from theorem 10   Q.E.D.

This is indeed a property of our abstract construction mechanism.

8   Streams in $\mathbf{Stream}_\sigma(\mathbf{Next})(X, X)$ as Monoid Homomorphisms

Next-based construction $\mathbf{Stream}_\sigma(\mathbf{Next})(X, X))$ that we consider is viewed as the category of relations on sets. In this context, we justify streams as *monoid homomorphisms*.

Note that the triples $(\mathbb{N}, +, 0)$ and $(\mathbb{N}, *, 1)$ are the monoids. Thus, as a convention, when we consider a monoid $\mathbb{N}$ that implies either. Apparently, there exist monoid homomorphisms between the monoid $\mathbb{N}$ and every monoid in $\mathbf{Stream}_\sigma(\mathbf{Next})(X, X)$.

Definition 20. *We define streams as the monoid homomorphisms between the monoid $\mathbb{N}$ and every monoid in $\mathbf{Stream}_\sigma(\mathbf{Next})(X, X)$.*

Formally, every stream is defined by the monoid homomorphism:

$$\mathbb{N}\xrightarrow{\;[i]\;}(1\xrightarrow{\;\mathbf{Next}_i(X,X)\;}Obj(\mathbf{Stream}_\sigma(\mathbf{Next})(X, X)))$$

where

$$Obj(\mathbf{Stream}_\sigma(\mathbf{Next})(X, X)) = \{\mathbf{Next}_i(X, X), \forall i \in T\}$$

In general, these monoid homomorphisms are drawn in the following commutative diagram.

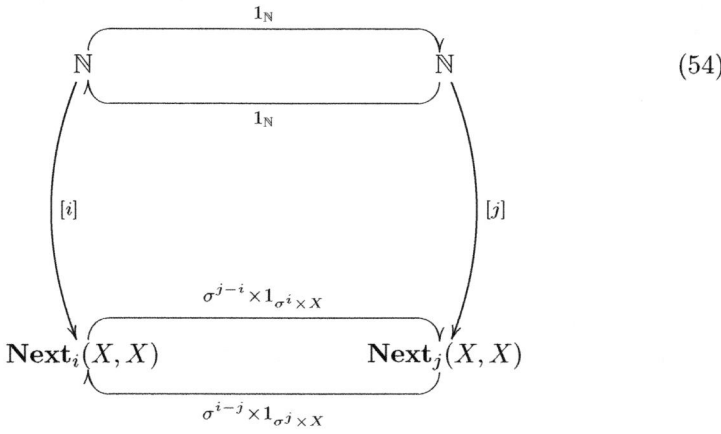

$$(54)$$

The monoid homomorphisms must satisfy the equations:

$$[0] = 1_X \tag{55}$$

and

$$1_{\mathbb{N}}; [j] = [i]; (\sigma^{j-i} \times 1_{\sigma^i \times X}) \tag{56}$$

or

$$1_{\mathbb{N}}; [i] = [j]; (\sigma^{i-j} \times 1_{\sigma^j \times X}) \tag{57}$$

By canceling identity, the equations (56) and (57) can be written such that

$$[j] = [i]; (\sigma^{j-i} \times 1_{\sigma^i \times X}) \tag{58}$$

or

$$[i] = [j]; (\sigma^{i-j} \times 1_{\sigma^j \times X}) \tag{59}$$

From the stream equations (55), (58) and (59) we are able to derive stream $[i]$ for every $i \in \mathbb{N}$. Derivation of streams is simplified by the following inference rules of categorical programming:

$$\overline{[0] = 1_X} \tag{60}$$

This means that there exists always an initial stream.

$$[0] \quad \text{and} \quad \mathbf{Next}(X, X) \xrightarrow{\sigma^i \times 1_X} \mathbf{Next}_i(X, X) \tag{61}$$

with $\mathbb{N}$, $\mathbf{Next}_i(X, X)$, and $[i] = \sigma^i \times 1_X$.

This rule states that given stream $[0]$ and isomorphism between two DASNs of $\mathbf{Next}(X, X)$ and $\mathbf{Next}_i(X, X)$ we can compute stream $[i]$ for every i.

$$[i] \quad \text{and} \quad \mathbf{Next}_i(X, X) \xrightarrow{\sigma^{j-i} \times 1_{\sigma^i \times X}} \mathbf{Next}_j(X, X) \tag{62}$$

with $\mathbb{N}$, $\mathbf{Next}_j(X, X)$, and $[j] = [i]; (\sigma^{j-i} \times 1_{\sigma^i \times X})$.

This rule means that given stream $[i]$ and isomorphism between two DASNs of $\mathbf{Next}_i(X, X)$ and $\mathbf{Next}_j(X, X)$ we can compute stream $[j]$ for every $j \neq i$.

From these three rules, we can prove

Theorem 12. *Every stream can be constructed by any existing stream*

Proof. It is straightforward by applying rule in (61) and/or rule in (62) Q.E.D.

This theorem simply says that every resulting stream can be shaped by any existing stream, it does not actually state that the resulting stream must be unique up to monoid homomorphism. However, we expect the uniqueness of every resulting stream and we can prove that this is indeed a property of our abstract construction mechanism. In fact, this gives rise to another theorem which shows the uniqueness to be true.

Theorem 13. *Every resulting stream, which is reached from any existing stream, is unique up to monoid homomorphism.*

Proof. Suppose that a resulting stream has two monoid homomorphisms $[i](\mathbb{N})$ and $[\widetilde{i}](\mathbb{N})$ which are formed by two distinct existing streams. The uniqueness of resulting stream as a unique monoid homomorphism is straightforwardly justified by the coinductive proof principle in (45) to result in $[i](\mathbb{N}) = [\widetilde{i}](\mathbb{N})$.

Let $\sim = \{\langle [i](\mathbb{N}), [\widetilde{i}](\mathbb{N})\rangle \mid \mathbb{N} \text{ is a monoid}\}$. We prove that $\sim$ is a bisimulation on $\mathbf{Next}_i(X, X)$. Consider $\langle [i](\mathbb{N}), [\widetilde{i}](\mathbb{N})\rangle \in \sim$, then

$$[i](\mathbb{N}) \text{ and } [\widetilde{i}](\mathbb{N}) \text{ are monoid homomorphims } \vdash \begin{pmatrix} 1 \overset{0}{\Longrightarrow} ([i](\mathbb{N})) \\ = \\ 1 \overset{0}{\Longrightarrow} ([\widetilde{i}](\mathbb{N})) \end{pmatrix}$$

$$[i](\mathbb{N}) \text{ and } [\widetilde{i}](\mathbb{N}) \text{ are monoid homomorphims } \vdash \begin{pmatrix} [i]^{\langle 1\rangle}(\mathbb{N}) = [i](\mathbb{N}^{\langle 1\rangle}) \\ \text{and} \\ [\widetilde{i}]^{\langle 1\rangle}(\mathbb{N}) = [\widetilde{i}](\mathbb{N}^{\langle 1\rangle}) \end{pmatrix}$$

$$\begin{pmatrix} [i]^{\langle 1\rangle}(\mathbb{N}) = [i](\mathbb{N}^{\langle 1\rangle}) \\ \text{and} \\ [\widetilde{i}]^{\langle 1\rangle}(\mathbb{N}) = [\widetilde{i}](\mathbb{N}^{\langle 1\rangle}) \\ \text{and} \\ \langle [i](\mathbb{N}^{\langle 1\rangle}), [\widetilde{i}](\mathbb{N}^{\langle 1\rangle})\rangle \in \sim \end{pmatrix} \vdash \langle [i]^{\langle 1\rangle}(\mathbb{N}), [\widetilde{i}]^{\langle 1\rangle}(\mathbb{N})\rangle \in \sim$$

$$\begin{pmatrix} \langle [i]^{\langle 1\rangle}(\mathbb{N}), [\widetilde{i}]^{\langle 1\rangle}(\mathbb{N})\rangle \in \sim \\ \text{and} \\ 1 \overset{0}{\Longrightarrow} ([i](\mathbb{N})) = 1 \overset{0}{\Longrightarrow} ([\widetilde{i}](\mathbb{N})) \end{pmatrix} \vdash \sim \text{ is a bisimulation}$$

In addition, for monoid $\mathbb{N}$, $\langle [i](\mathbb{N}), [\widetilde{i}](\mathbb{N})\rangle \in \sim$ iff $[i](\mathbb{N}) \sim [\widetilde{i}](\mathbb{N})$. Thus, by the coinductive proof principle in (45) then $[i] = [\widetilde{i}]$.   Q.E.D.

9   Generalizing $\mathbf{Stream}_\sigma(\mathbf{Next})(X, X)$

In this section, we construct $\mathbf{Stream}_\sigma(\mathbf{Next})(\_, \_)$ as a generalized $\mathbf{Stream}_\sigma(\mathbf{Next})(X, X)$. In fact, for any pair of objects X, Y, we define

$$Obj(\mathbf{Stream}_\sigma(\mathbf{Next})(\_, \_)) = \{\mathbf{Next}_i(X, Y), \forall i \in \mathbb{N}\}$$

Morphism between two objects $\mathbf{Next}_i(X, Y)$ and $\mathbf{Next}_j(Y, Z)$ in $\mathbf{Stream}_\sigma(\mathbf{Next})(\_, \_)$ is defined by a composition "$;$" such that if f is in $\mathbf{Next}_i(X, Y)$ and g in $\mathbf{Next}_j(Y, Z)$ then

$$f; g = (1_{\sigma^j} \times f) \times (1_{\sigma^i} \times g)$$

$$f;g=(1_{\sigma^j}\times f)\times(1_{\sigma^i}\times g)$$

In other words, $1 \xrightarrow{f} \mathbf{Next}_i(X,Y) \qquad 1 \xrightarrow{g} \mathbf{Next}_j(Y,Z)$. It follows that

Theorem 14. Stream$_\sigma$(Next)($\_,\_$) *defines a category*

Proof. For all f in $\mathbf{Next}_i(X,Y)$ and g in $\mathbf{Next}_j(Y,Z)$, the composition ";" exists such that

$$\vdash_{def} f \in \mathbf{Next}_i(X,Y) \tag{63}$$

$$(63) \quad \vdash (1_{\sigma^j} \times f) \in \mathbf{Next}_{i+j}(X,Y) \tag{64}$$

$$\vdash_{def} g \in \mathbf{Next}_j(Y,Z) \tag{65}$$

$$(65) \quad \vdash (1_{\sigma^i} \times g) \in \mathbf{Next}_{i+j}(Y,Z) \tag{66}$$

By (64) and (66), the composition $f;g$ is completely defined and a member of $\mathbf{Next}_{i+j}(X,Z)$. That is

$$1 \xrightarrow{f} \mathbf{Next}_i(X,Y) \qquad 1 \xrightarrow{g} \mathbf{Next}_j(Y,Z) \equiv 1 \xrightarrow{f;g} \mathbf{Next}_{i+j}(X,Z)$$

Associativity of ";" follows immediately from the associativity of $\times$ in every **Next**.

Finally, we consider two distinct cases as follows:

Firstly, when $i = 0$. For every morphism $\mathbf{Next}(X,Y) \xrightarrow{f;g} \mathbf{Next}(Y,Z)$, that is $(f;g)$ $\in$ $\mathbf{Next}(X,Z)$, the identities are specified by $\mathbf{Next}(X,X) \xrightarrow{1_X} \mathbf{Next}(X,X)$ and $\mathbf{Next}(Z,Z) \xrightarrow{1_Z} \mathbf{Next}(Z,Z)$ such that the following equation holds: $1_X;(f;g) = (f;g) = (f;g);1_Z$.

Secondly, when $i > 0$. The identities are also consistent with every morphism $\mathbf{Next}_i(X,Y) \xrightarrow{f;g} \mathbf{Next}_j(Y,Z)$, that is $(f;g) \in \mathbf{Next}_{i+j}(X,Z)$. In fact, if $(f;g) \in \mathbf{Next}_{i+j}(X,Z)$ then $(f;g) \in \mathbf{Next}(X',Z')$ with $X' = \sigma^{i+j} \times X$ and $Z' = \sigma^{i+j} \times Z$. Thus, the following equation holds: $1_{X'};(f;g) = (f;g) = (f;g);1_{Z'}$ Q.E.D.

10   Conclusions

In this paper, using categorical language, data aware computing in self-organizing networks such as wireless sensor networks (WSNs), mobile ad hoc networks

(MANETs) and so on has categorically been interpreted. In other words, a foundation of data aware computing in such self-organizing networks has formally been established. We have constructed some algebraic models for data aware self-organizing networks (DASNs) where all the algebraic models are based on mapping a DASN configuration to another. In the specification, we have considered the DASN configuration at every reorganization step to be a member of the set $\sigma^i \times C$. We have developed $C \xrightarrow{ds} \sigma$ as a model of data streams in DASNs and, by the way, quantitative behaviors $C \xrightarrow{B(\_)} \mathbb{R}$ of the data streams have symbolically been evaluated. Every $\mathbf{Next}_i(X, X)$ has been constructed as monoids of data aware self-organizations to shape streams of data aware self-organizations. By the monoids $\mathbf{Next}_i(X, X)$, we have formed $\mathbf{Stream}_\sigma(\mathbf{Next})(X, X)$ to be a category of data aware self-organizations monoids. $\mathbf{Next}$-based construction of $\mathbf{Stream}_\sigma(\mathbf{Next})(X, X))$ under the investigation is viewed as the category of relations on sets where the streams have been justified as monoid homomorphisms since there exist monoid homomorphisms between the monoid $\mathbb{N}$ and every monoid in $\mathbf{Stream}_\sigma(\mathbf{Next})(X, X)$. Category $\mathbf{Stream}_\sigma(\mathbf{Next})(X, X)$ of the monoids has also been generalized to be category $\mathbf{Stream}_\sigma(\mathbf{Next})(X, Y)$ as a more extended view on the considered topic.

References

1. Adamek, J., Herrlich, H., Strecker, G.: Abstract and Concrete Categories. John Wiley and Sons (1990)
2. Asperti, A., Longo, G.: Categories, Types and Structures. M.I.T. Press (1991)
3. Bergman, G.M.: An Invitation to General Algebra and Universal Constructions. Henry Helson, 15 the Crescent, Berkeley CA, US (1998)
4. Bowen, J.P., Hinchey, M.G.: Formal Methods. In: Tucker Jr., A.B. (ed.) Computer Science Handbook, ch. 106, 2nd edn. Section XI, Software Engineering, pp. 106-1–106-25. Chapman & Hall / CRC, ACM, USA (2004)
5. Butera, W.: Text Display and Graphics Control on a Paintable Computer. In: Serugendo, G.D.M., Flatin, J.P.M., Jelasity, M. (eds.) Proceedings of 1st International Conference on Self-Adaptive and Self-Organizing Systems (SASO 2007), Boston, Massachusetts, USA, July 9-11, pp. 45–54. IEEE Computer Society Press (2007)
6. Calisti, M., Meer, S.V.D., Strassner, J.(eds.): Advanced Autonomic Networking and Communication. Whitestein Series in Software Agent Technologies and Autonomic Computing, 190 pages. Springer, Heidelberg (2008)
7. Herz, M., Hartenstein, R., Miranda, M., Brockmeyer, E., Catthoor, F.: Memory Addressing Organization for Stream-based Reconfigurable Computing. In: Proceedings of 9th International Conference on Electronics, Circuits and Systems (ICECS), Dubrovnik, Croatia, September 15-18, vol. 2, pp. 813–817. IEEE (2002)
8. Heylighen, F.: The science of self-organization and adaptivity. In: Kiel, L.D. (ed.) The Encyclopedia of Life Support Systems, Knowledge Management, Organizational Intelligence and Learning, and Complexity. EOLSS Publishers, Oxford (2002), Retrieved from, http://www.eolss.net
9. Hinchey, M.G., Bowen, J.P.: Industrial-Strength Formal Methods in Practice. In: Formal Approaches to Computing and Information Technology (FACIT). Springer, London (1999)

10. Jacobs, B., Rutten, J.: A Tutorial on (Co)Algebras and (Co)Induction. Bulletin of EATCS 62, 222–259 (1997)
11. Jin, X., Liu, J.: From Individual Based Modeling to Autonomy Oriented Computation. In: Nickles, M., Rovatsos, M., Weiss, G. (eds.) AUTONOMY 2003. LNCS (LNAI), vol. 2969, pp. 151–169. Springer, Heidelberg (2004)
12. Ko, S., Gupta, I., Jo, Y.: Novel Mathematics-Inspired Algorithms for Self-Adaptive Peer-to-Peer Computing. In: Serugendo, G.D.M., Flatin, J.P.M., Jelasity, M. (eds.) Proceedings of 1st International Conference on Self-Adaptive and Self-Organizing Systems (SASO 2007), Boston, Massachusetts, USA, July 9-11, pp. 3–12. IEEE Computer Society Press (2007)
13. Li, J.: On peer-to-peer (P2P) content delivery. Peer-to-Peer Networking and Applications 1(1), 45–63 (2008)
14. Pacheco, O.: Autonomy in an Organizational Context. In: Nickles, M., Rovatsos, M., Weiss, G. (eds.) AUTONOMY 2003. LNCS (LNAI), vol. 2969, pp. 195–208. Springer, Heidelberg (2004)
15. Parashar, M., Hariri, S.: Autonomic Computing: Concepts, Infrastructure and Applications, 1st edn., 568 pages. CRC Press (December 2006)
16. Prokopenko, M.: Design vs. Self-organization. In: Advances in Applied Self-organizing Systems, 1st edn. Advanced Information and Knowledge Processing, pp. 3–18. Springer, Heidelberg (2008)
17. Rutten, J.J.M.M.: Universal Coalgebra: A Theory of Systems. Theoretical Computer Science 249(1), 3–80 (2000)
18. Rutten, J.J.M.M.: Elements of Stream Calculus (An Extensive Exercise in Coinduction). Electronic Notes in Theoretical Computer Science, vol. 45. Elsevier Science Publishers Ltd. (2001)
19. Rutten, J.J.M.M.: Algebra, Bitstreams, and Circuits. Technical Report SEN-R0502, CWI, Amsterdam, The Netherlands (2005)
20. SASO. Manifesto on Self-organizing Systems. Retrieved from (2008), http://polaris.ing.unimo.it/saso2008/
21. Vinh, P.C.: Formal Aspects of Dynamic Reconfigurability in Reconfigurable Computing Systems. PhD thesis, London South Bank University, 103 Borough Road, London SE1 0AA, UK, May 4 (2006)
22. Vinh, P.C.: Homomorphism between AOMRC and Hoare Model of Deterministic Reconfiguration Processes in Reconfigurable Computing Systems. Scientific Annals of Computer Science XVII, 113–145 (2007)
23. Vinh, P.C.: Formal Aspects of Self-* in Autonomic Networked Computing Systems. In: Autonomic Computing and Networking, 1st edn. Springer, USA (2009)
24. Vinh, P.C.: Categorical Approaches to Models and Behaviors of Autonomic Agent Systems. The International Journal of Cognitive Informatics and Natural Intelligence (IJCiNi) 3(1), 17–33 (2009)
25. Vinh, P.C.: Formalizing Parallel Programming in Large Scale Distributed Networks: From Tasks Parallel and Data Parallel to Applied Categorical Structures. In: Parallel Programming and Applications in Grid, P2P and Network-based Systems, 1st edn. Advances in Parallel Computing. IOS Press (2009)
26. Vinh, P.C., Bowen, J.P.: Formalising Configuration Relocation Behaviours for Reconfigurable Computing. In: Proceedings of SFP Workshop, FDL 2002: Forum on Specification & Design Languages, Marseille, France, September 24-27, CD-ROM (2002)

27. Vinh, P.C., Bowen, J.P.: An Algorithmic Approach by Heuristics to Dynamical Reconfiguration of Logic Resources on Reconfigurable FPGAs. In: Proceedings of 12th International Symposium on Field Programmable Gate Arrays, Monterey, CA, USA, February 22-24, page 254. ACM/SIGDA (2004)

28. Vinh, P.C., Bowen, J.P.: A Provable Algorithm for Reconfiguration in Embedded Reconfigurable Computing. In: Hinchey, M.G. (ed.) Proceedings of 29th Annual IEEE/NASA Software Engineering Workshop (SEW), Greenbelt, MD, USA, April 6-7, pp. 245–252. IEEE Computer Society Press (2005)

29. Vinh, P.C., Bowen, J.P.: Continuity Aspects of Embedded Reconfigurable Computing. Innovations in Systems and Software Engineering: A NASA Journal 1(1), 41–53 (2005)

30. Vinh, P.C., Bowen, J.P.: POM Based Semantics of RTL and a Validation Method for the Synthesis Results in Dynamically Reconfigurable Computing Systems. In: Rozenblit, J., O'Neill, T., Peng, J. (eds.) Proceedings of 12th Annual International Conference and Workshop on the Engineering of Computer Based Systems (ECBS), Greenbelt, MD, USA, April 4-5, pp. 247–254. IEEE Computer Society Press (2005)

31. Vinh, P.C., Bowen, J.P.: A Formal Approach to Aspect-Oriented Modular Reconfigurable Computing. In: Proceedings of 1st IEEE & IFIP International Symposium on Theoretical Aspects of Software Engineering (TASE), Shanghai, China, June 6-8, pp. 369–378. IEEE Computer Society Press (2007)

32. Vinh, P.C., Bowen, J.P.: Formalization of Data Flow Computing and a Coinductive Approach to Verifying Flowware Synthesis. In: Gavrilova, M.L., Tan, C.J.K. (eds.) Transactions on Computational Science I. LNCS, vol. 4750, pp. 1–36. Springer, Heidelberg (2008)

33. Wang, Y.: Toward Theoretical Foundations of Autonomic Computing. The International Journal of Cognitive Informatics and Natural Intelligence (IJCiNi) 1(3), 1–16 (2007)

34. Witkowski, M., Stathis, K.: A Dialectic Architecture for Computational Autonomy. In: Nickles, M., Rovatsos, M., Weiss, G. (eds.) AUTONOMY 2003. LNCS (LNAI), vol. 2969, pp. 261–273. Springer, Heidelberg (2004)

35. Wolf, T.D., Holvoet, T.: A Taxonomy for Self-* Properties in Decentralized Autonomic Computing. In: Autonomic Computing: Concepts, Infrastructure and Applications, 1st edn., pp. 101–120. CRC Press (2006)

36. Yang, B., Liu, J.: An Autonomy Oriented Computing (AOC) Approach to Distributed Network Community Mining. In: Serugendo, G.D.M., Flatin, J.P.M., Jelasity, M. (eds.) Proceedings of 1st International Conference on Self-Adaptive and Self-Organizing Systems (SASO 2007), Boston, Massachusetts, USA, July 9-11, pp. 151–160. IEEE Computer Society Press (2007)

Accelerated Evolution: A Biologically-Inspired Approach for Augmenting Self-star Properties in Wireless Sensor Networks

Pruet Boonma[1] and Junichi Suzuki[2]

[1] Department of Computer Engineering
Chiang Mai University
pruet@eng.cmu.ac.th
[2] Department of Computer Science
University of Massachusetts, Boston
jxs@cs.umb.edu

Abstract. Wireless sensor networks (WSNs) possess inherent tradeoffs among conflicting performance objectives such as data yield, data fidelity and power consumption. In order to address this challenge, this paper proposes a biologically-inspired application framework for WSNs. The proposed framework, called El Niño, models an application as a decentralized group of software agents. This is analogous to a bee colony (application) consisting of bees (agents). Agents collect sensor data on individual nodes and carry the data to base stations. They perform this data collection functionality by autonomously sensing their local network conditions and adaptively invoking biological behaviors such as pheromone emission, swarming, reproduction and migration. Each agent carries its own operational parameters, as genes, which govern its behavior invocation and configure its underlying sensor nodes. El Niño allows agents to evolve and adapt their operational parameters to network dynamics and disruptions by seeking the optimal tradeoffs among conflicting performance objectives. This evolution process is augmented by a notion of accelerated evolution. It allows agents to evolve their operational parameters by learning dynamic network conditions in the network and approximating their performance under the conditions. This is intended to expedite agent evolution to adapt to network dynamics and disruptions.

1 Introduction

Wireless sensor networks (WSNs) have inherent tradeoffs among conflicting performance objectives such as data yield, data fidelity and power efficiency [1,2]. For example, hop-by-hop recovery is often applied for packet transmission in order to improve data yield (the quantity of collected data). However, this can degrade data fidelity (the quality of collected data; e.g., data freshness). For improving data fidelity, sensor nodes may transmit data to base stations through the shortest paths; however, data yield can degrade because of traffic congestion and packet losses on the paths.

Performance objectives tend to conflict because WSNs possess a number of operational parameters and a single parameter often impacts multiple performance objectives simultaneously. For example, per-node sleep period impacts data fidelity and power efficiency; data fidelity degrades but power efficiency improves by increasing sleep period.

M.L. Gavrilova et al. (Eds.): Trans. on Comput. Sci. XV, LNCS 7050, pp. 108–129, 2012.

Moreover, there exist conflicts among operational parameters. This makes performance objectives conflict severely. For example, transmission timeout period and sleep period conflict with each other. Increasing timeout period can improve power efficiency because it reduces the number of data retransmissions. However, it can decrease sleep period, which in turn degrades power efficiency.

In order to address this challenge, the authors of the paper envision autonomic WSN applications that understand their conflicting performance objectives, find the optimal tradeoffs among the objectives to tune operational parameters, and autonomously adapt to network dynamics and disruptions such as node/link failures. For making this vision a reality, this paper proposes an autonomic sensor networking framework, El Niño, which allows WSN applications to exhibit the following properties:

- *Self-configuration*: allows WSN applications to configure their own operational parameters and self-organize[1] into desirable structures and patterns (e.g., routing structures and duty cycling patterns).
- *Self-optimization*: allows WSN applications to constantly seek improvement in their performance by adapting to network dynamics with minimal human intervention.
- *Self-healing*: allows WSN applications to autonomously detect and recover from disruptions in the network (e.g., node and link failures).

As an inspiration for the design strategy of El Niño, the authors of the paper observe that various biological systems have developed the mechanisms necessary to realize the vision of El Niño. For example, a bee colony self-organizes to satisfy conflicting objectives simultaneously for maintaining its well-being [4]. Those objectives include maximizing the amount of collected honey, maintaining the temperature in a nest and minimizing the number of dead drones. If bees focus only on foraging, they fail to ventilate their nest and remove dead drones. Given this observation, El Niño applies key biological mechanisms to implement autonomic WSN applications.

Figure 1 shows the architecture of El Niño. The El Niño runtime operates atop TinyOS [5] on each node. It consists of two types of software components: *agents* and *middleware platforms*, which are modeled after bees and flowers, respectively. Each application is designed as a decentralized group of agents. This is analogous to a bee colony (application) consisting of bees (agents). Agents collect sensor data on platforms (flowers) on nodes, and carry the data to base stations on a hop-by-hop basis, in turn, to the El Niño server (Figure 1), which is modeled after a nest of bees.

Agents perform this data collection functionality by autonomously sensing their local and surrounding network conditions (e.g., network traffic and node/link failures) and adaptively invoking biological behaviors such as pheromone emission, swarming, reproduction, migration and death. A middleware platform runs on each node and hosts one or more agents (Figure 1). It provides a series of runtime services that agents use to perform their data collection functionalities and behaviors.

---

[1] Self-organization is a process in which a system's internal components autonomously react to environmental changes, interact with each other and create an ordered state without being guided by any outside sources [3].

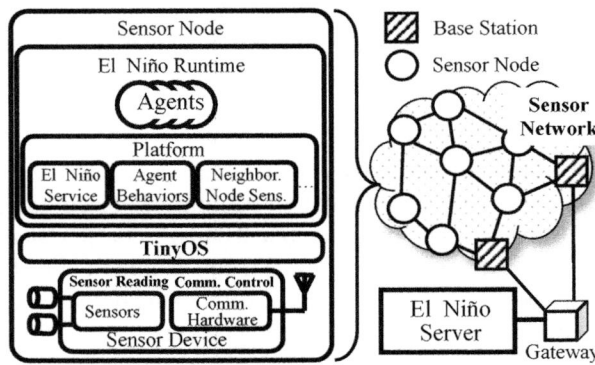

Fig. 1. The Architecture of El Niño

El Niño implements an evolutionary multiobjective optimization algorithm for agents. Each agent possesses operational parameters, as *genes*, which govern its behavior invocation and configure its underlying nodes. El Niño allows agents to evolve their genes (i.e., operational parameters) via genetic operations (e.g., crossover, mutation and selection) and adapt them to dynamic network conditions by seeking the optimal tradeoffs among conflicting performance objectives.

This evolution process frees application designers from anticipating all possible network conditions and tuning their agents' operational parameters to the conditions at design time. Instead, agents evolve and autonomously adapt their parameters at runtime. This can significantly simplify the implementation of agents (i.e., WSN applications).

El Niño implements a notion of *accelerated evolution* as well as a regular evolution process described above. In the regular evolution, agents use their operational parameters for their data collection in the network and then evolve their parameters based on their performance results. With accelerated evolution, agents can evolve their parameters by learning dynamic network conditions and approximating their performance under the conditions without actually using the parameters in the network. This way, accelerated evolution allows agents to expedite their evolution and efficiently adapt to network dynamics and disruptions.

This paper describes the design of El Niño and evaluates it through simulations. Simulation results show that El Niño allows agents to exhibit self-configuration, self-optimization and self-healing properties in dynamic networks by evolving their operational parameters with respect to conflicting performance objectives. El Niño's accelerated evolution allows agents to augment their self-configuration, self-optimization and self-healing abilities by gaining significant (up to 43.3%) improvement in performance convergence.

2   Agents in El Niño

El Niño is currently designed to implement data collection applications. An agent is initially deployed with a randomly-generated operational parameters on each node. Each agent collects sensor data on a node at each duty cycle and carries the data to a base station on a hop-by-hop basis. Each agent invokes six behaviors in the following order at each duty cycle.

Step 1: Energy gain. Each agent collects sensor data and gain *energy*. The concept of energy does not represent the amount of physical batteries in a node. Instead, it is a logical concept that impacts agent behaviors. Each agent updates its energy level with a constant energy intake at every duty cycle.

Step 2: Energy expenditure and death. Each agent consumes a constant amount of energy to use computing/networking resources available on a node (e.g., CPU and radio transmitter). It also expends energy to invoke its behaviors. The energy costs to invoke behaviors are constant for all agents. An agent dies when its energy level becomes zero. The death behavior is intended to eliminate the agents that have ineffective operational parameters. For example, an agent would die before arriving at a base station if it follows a too long migration path. When an agent dies, the local platform removes the agent and releases all resources allocated to it[2].

Step 3: Replication. Each agent makes n_c copies of itself. Replicated agents are placed on the node that their parent resides on, and they inherit the parent's operational parameters as well as a constant amount of energy. Mutation may occur at the probability of $1/n_b$ to randomly alter each of the inherited operational parameters. (n_b denotes the number of operational parameters that each agent has. Probabilistically, one operational parameter is altered via mutation.) Each child agent contains the sensor data that its parent has collected, and carries it to a base station. Different child agents may choose different paths to a base station depending on their operational parameters.

Step 4: Swarming. Each agent may swarm (or merge) with others at an intermediate node on its way to a base station. On each intermediate node, it waits for a particular period (t_w) for other agents to arrive at the node. If it meets the agents migrating to the same base station, it merges with them and aggregates the sensor data they carry. It also uses the operational parameters of the best one in those swarming/aggregating agents in terms of performance objectives. (See Section 3 on how to determine the best-performing agent.) The swarming behavior is intended to save power consumption by aggregating multiple agents and reducing the number of data transmissions. If the size of data an agent carries exceeds the maximum size of a packet, the agent does not perform the swarming behavior.

Step 5: Pheromone sensing and migration. On each intermediate node toward a base station, each agent chooses the next-hop node in its migration by sensing three types of *pheromones* available on the local node: base station, migration and alert pheromones.

Each base station periodically propagates a base station pheromone to individual nodes. Their concentration decays on a hop-by-hop basis. Using base station pheromones on each neighboring nodes, agents can sense where base stations exist approximately, and move toward them by climbing a concentration gradient of base station pheromones.

Agents emit migration pheromones on their local nodes when they migrate to neighboring nodes. Each migration pheromone references the destination node an agent has migrated to. Agents also emit alert pheromones when they fail migrations within a

---

[2] If all agents are dying on a node at the same time, a randomly selected one will survive.

timeout period. Migration failures can occur due to, for example, node/link failures. Each alert pheromone references the node that an agent could not migrate to. Each of migration and alert pheromones has its own concentration, which decays by half at each duty cycle. A pheromone disappears when its concentration becomes zero.

Each agent examines Equation 1 to determine which next-hop node it migrates to.

$$S_j = \sum_{t=1}^{3} w_t \frac{P_{t,j} - P_{t_{min}}}{P_{t_{max}} - P_{t_{min}}} \tag{1}$$

An agent calculates this weighted sum (S_j) for each neighboring node j, and moves to a node that generates the highest weighted sum. t denotes pheromone type; $P_{1,j}$, $P_{2,j}$ and $P_{3,j}$ represent the concentrations of base station, migration and alert pheromones on the node j. $P_{t_{max}}$ and $P_{t_{min}}$ denote the maximum and minimum concentrations of P_t among all neighboring nodes.

The weight values in Equation 1 ($w_t, 1 \leq t \leq 3$) govern how agents perform the migration behavior. For example, if an agent has zero for w_2 and w_3, it ignores migration and alert pheromones, and moves toward a base station by climbing a concentration gradient of base station pheromones. If an agent has a positive value for w_2, it follows a migration pheromone trace on which many other agents have traveled. The trace can be the shortest path to a base station. Conversely, a negative w_2 value allows an agent to go off a migration pheromone trace and follow another path to a base station. This helps avoid separating the network into islands. The network can be separated with the migration paths that too many agents follow because the nodes on the paths run out of their battery earlier than the others. If an agent has a negative value for w_3, it avoids moving to a node referenced by an alert pheromone, thereby bypassing failed nodes and links.

Step 6: Pheromone emission. When an agent migrates, it emits a migration pheromone on the local node. If its migration fails, it emits an alert pheromone on the local node. Each alert pheromone spreads to direct neighbor nodes.

3   Evolutionary Multiobjective Optimization with El Niño

This section describes operational parameters, performance objectives, evolution process and accelerated evolution in El Niño.

3.1   Operational Parameters

Each agent carries its own operational parameters, as genes, which govern its behavior invocations and configure a node that it resides on. El Niño currently considers twelve parameters. There are two types of operational parameters: agent behavior policy parameters and node configuration parameters.

An agent behavior policy consists of five parameters: the number of copies an agent makes in each duty cycle (n_c), waiting period in the swarming behavior (t_w) and three weight values in Equation 1 (w_t, $1 \leq t \leq 3$). n_c and t_w are non-negative. Parameters in an agent's behavior policy govern its behavior invocations.

Node configuration parameters are used to configure the operations of a node. If multiple agents run on a node, the best-performing agent among them configures the node with its node configuration parameters. See Section 3 on how to determine the best-performing agent with respect to given performance objectives.)

Transmission timeout period indicates the time period that a node waits after it transmits a data (e.g, for agent migration or pheromone emission). Higher transmission time out allows sensor node to be more tolerance to the delay in data communication among sensor nodes. However, this might delays the agent retransmission when the agent actually lost. Lower transmission time out reduce the time delays for transmission recovery, but it might cause duplicate transmission.

Maximum number of retransmissions per duty cycle dictate the number of retransmission that a sensor node can perform in each duty cycle. When a transmission fails, sensor node tries to retransmit the data to do hop-by-hop recovery. This parameter controls how many time sensor node keeps trying for each data (e.g., for each agent migration or each pheromone emission). Higher number of retransmission allows agents/pheromone to successfully transmitted in a noisy communication channel environment. However, this might cause a huge power consumption from agents with inferior behavior policy, e.g., agents keep trying to migrate to death sensor nodes.

Maximum number of data transmission per duty cycle controls the maximum number of total transmission a sensor node can perform in each duty cycle. Higher value of data transmission per duty cycle allows sensor node to send out more data in each duty cycle; however, this might introduce too much traffic in the network channel which leads to congestion. Lower value of this parameter might reduce the congestion in the network channel but this might delay data transmission from sensor nodes.

Queue size for agents governs the number of agents that a sensor node can maintain at a time. Sensor nodes start dropping incoming agents when the number of local agents is higher than the queue size. Increasing this number allows more agents to stay local, but this might increase the outgoing traffic from the sensor node.

Sensor node parameters impact the characteristic of sensor node that agents are located. Again, if there are multiple agents in a same sensor node, the network parameters from the gene of the best agent will be used.

Sleep period is the period for which a node sleeps between two duty cycles. The longer a node sleeps, the less amount of power it consumes. However, it can degrade latency because a sleeping node receives and sends out no agents nor pheromones.

Sensing rate is a multiple of sleep period. The shorter the sensing rate, the higher amount of data reporting to a base station. Higher amount of data increases the data accuracy. However, it can increases the power consumption of sensor nodes.

Reporting rate is a multiple of sensing period. Sensor nodes collect sensor data in each sensing rate and report them to a base station according to reporting rate, e.g., if reporting rate is two, sensor nodes collect two data and report them at once. The longer the reporting rate, the lower power the sensor node consume. However, it can decrease the data accuracy.

3.2 Performance Objectives

Each agent considers six conflicting performance objectives related to data yield, data fidelity and power consumption: latency, success rate, cost, the degree of data aggregation, sleep period and data accuracy. Success rate and the degree of data aggregation impact data yield. Latency and data accuracy impact date fidelity. Cost and sleep period impact power consumption. El Niño strives to minimize latency, cost and sleep period and maximize success rate, the degree of data aggregation and data accuracy.

(1) **Latency** (L) indicates the time required for an agent to travel to a base station from a node where the agent is replicated. It is measured as the ratio of this travel time (t) to the physical distance between the base station and a node where the agent is replicated (d); $L = \frac{t}{d}$. The El Niño server knows each node's location with a certain localization method.

(2) **Success rate** (S) is measured as the ratio of the number of sensor data carried to base stations (n_{arrive}) to the total number of nodes in the network (N); $S = n_{arrive}/N$.

(3) **Cost** (C) represents power consumption required for an agent to travel to a base station:

$$C = \frac{(E_{sense} \times R_{sense}) + R_{report} \times (N_{tx} \times (E_{tx} \times S_{data} + E_{radio}))/d}{N_{data}} \tag{2}$$

E_{sense}, E_{tx} and E_{radio} denote power consumption (in mW) to collect data from a sensor, transmit a bit of data between two neighboring nodes and operate the radio circuit in a node. These constants are obtained from [6]. R_{sense} and R_{report} denote an agent's sensing rate and reporting rate. N_{tx} is the total number of node-to-node transmissions for an agent to arrive at a base station. This counts successful and failed migrations of an agent as well as the transmissions of its migration and alert pheromones. S_{data} denotes the size of an agent in bit. N_{data} denotes the total number of data carried by an agent.

(4) **Degree of data aggregation** is measured as the number of sensor data in an agent. It is more than one in a swarming agent. The larger it is, the less amount of power agents consume because of a less number of migrations in the network. However, this can degrade latency because agents wait for others to swarm on nodes.

(5) **Sleep period** is the period for which a node sleeps between two duty cycles. The longer it is, the less amount of power it consumes. However, this can degrade latency because a sleeping node receives and transmits no agents nor pheromones. It can also increase power consumption of the other nodes if they re-transmit agents to the sleeping node.

(6) **Data accuracy** (A) is measured based on the mean squared error between collected sensor data (s) and estimated sensor data ($\hat{s}$).

$$A = \left(\frac{1}{n_{arrive}} \sum_{j=1}^{n_{arrive}} (s_j - \hat{s}_j)^2\right)^{-1} \tag{3}$$

$\hat{s}$ is calculated with a data prediction model using Autoregressive Integrated Moving Average (ARIMA). Higher data accuracy indicates higher quality in estimated data.

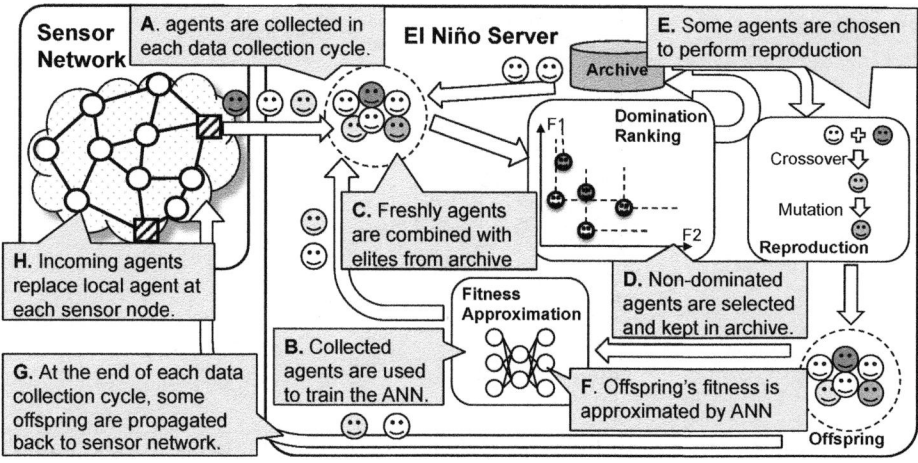

Fig. 2. An Overview of El Niño Server Process

3.3 El Niño Server Process

The optimization process in El Niño is performed in the El Niño Server (see Figure 1). In each data collection cycle, the elite selection process evaluates the agents that arrive at base stations, based on given performance objectives, and chooses the best (or elite) ones. Elite agents performs genetic operations (crossover and mutation) to produce next generation agents (offspring) which are propagated to individual nodes in the network. A next generation agent inherits operational parameters (genes) from its parents via crossover, and mutation may occur on the gene's operational parameter.

Reproduction is intended to evolve agents so that the agents that fit better to the environment become more abundant in the network. It retains the agents with a high fitness to the current network conditions (i.e., agents that have effective operational parameters, such as moving toward a base station in a short latency). It also eliminates the agents with a low fitness (i.e., agents that have ineffective operational parameters, such as consuming too much power to reach a base station). Through successive generations, effective operational parameters become abundant in a population of agents while ineffective ones become dormant or extinct. This allows agents to adapt to dynamic network conditions.

In between data collection cycle, additional offspring are produced from genetic operation and evaluated using artificial neural network. This operation intends to improve the quality of agents without the need to actual evaluating agents in the sensor network. As a result, agents quality improvement can be accelerated while the cost (e.g., sensor node's power consumption) to improve agents can be reduced.

Figure 2 shows the overview of the evolutionary algorithm performed at the El Niño Server. In each data collection cycle, the El Niño server collects agents arriving from sensor network (step A in Figure 2). The first step is to measure six objective values of each agent that arrives at base stations. Then, the incoming agents genes and

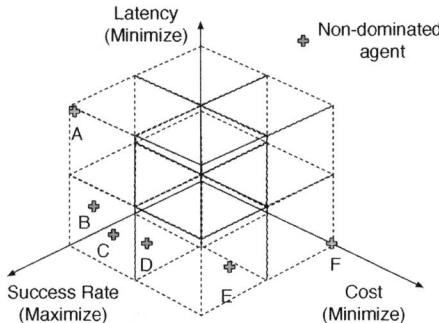

Fig. 3. An Example Elite Selection

objective values are used to train the artificial neural network (step B). The training process allows artificial neural network to adapt to the changed in network condition. The collected agents are combined with elite agents from archive and they are evaluated using domination ranking (step C and D). The dominated agents are eliminated and the non-dominated agents are preserved in the archive.

In El Niño, offspring are produced with two processes. First, offspring are produced at the end of data collection cycle and propagated to the sensor network to replace the agents at sensor nodes (step E., G. and H. in Figure 2). Second, in between data collection cycle, offspring may be produced and evaluated by artificial neural network before placed in the population (step E and F).

3.4   Offsprings Production for Each Data Collection Cycle

In both situations, a subset of non-dominated agents is selected as elite agents (step E.). This is performed with the objective space; a six dimensional hypercube space whose axes represent six objectives. Each axis is divided between the maximum and minimum objective values of non-dominated agents so that the space contains small cubes. Each non-dominated agent is plotted in the objective space based on their objective values. If multiple agents are plotted in the same cube, a single agent is randomly selected as an elite agent. If no agents are plotted in a cube, no elite agent is selected from the cube. This hypercube-based elite selection is designed to maintain the diversity of elite agents, which is can improve their adaptability even to unanticipated network conditions.

Figure 3 shows an example hypercube that shows three objectives (success rate, cost and latency). It is divided to two ranges for each objective; eight cubes exist in total. Thus, the maximum number of elite agents is eight. In this example, six (A to F) non-dominated agents are plotted. Three agents (B, C, and D) are plotted in the lower left cube, while the other three agents (A, E, and F) are plotted in three different cubes. From the lower left cube, only one agent is randomly selected as an elite agent. A, E, and F are selected as elite agents because they are in different cubes.

Consequently, the elites agents perform reproduction with randomly selected agent from archive and an offspring may perform mutation. Mutation rate is adjusted based on the disorderliness of the current population in the objective space. If the current population is very disordered, it indicates that the population does find good solutions yet. In the other words, the current population is still far from Pareto front or unable to form Pareto front yet. Therefore, mutation rate should be keep high in order to allow agents in the population to search more. However, if a population maintains a low level of disorder, which can indicates that agents in the population already form a Pareto front or very close to the Pareto front, mutation rate should be decreased to reduce the fluctuations in the objective space. El Niño measures the disorderliness of the current population using entropy of the hypercube (see Figure 3). All agents, including dominated or non-dominated ones, are plotted on the hypercube. Then, entropy H of the hypercube is measured as:

$$H = -\sum_{i \in C} P(i) \log_2(P(i)). \text{ where } P(i) = \frac{n_i}{\sum_{i \in C} n_i} \tag{4}$$

C denotes a set of cubes in a hypercube, and $P(i)$ denotes the probability that individuals exist in the i-th cube. n_i denotes the number of individuals in the i-th cube.

Mutation rate (m) is adjusted with normalized entropy (H_o):

$$m = m_{\max} \times \sqrt{1 - (1 - H_o)^2} \text{ where } H_o = \frac{H}{H_{\max}} = \frac{H}{\log_2 n}. \tag{5}$$

n denotes the total number of cubes in a hypercube. $m_{\max}$ denotes the maximum mutation rate.

At the end of data collection cycle, the offspring are propagated back to the sensor network (step G). When they arrive at each sensor node, one of propagated elite agents are selected based on its gene similarity with the local agent. Gene similarity is measured with the Euclidean distance between the values of two genes. If two or more elite agents have the same similarity to the local agent, one of them is randomly selected. The selected elite agent then replace the local agent (step H).

3.5 Offsprings Production in between Data Collection Cycle

In between data collection cycle, offspring may be produced and evaluated using artificial neural network (step E and F in Figure 2). The number of sub-generation to perform this reproduction is adaptively determined from the accuracy of the artificial neural network. When the collected agents (at step B) are used to train the artificial neural network, the accuracy of the artificial neural network is determined by measuring the average mean square error (E) of the predicted objective values and actual objective values of every collected agents:

$$E = \frac{1}{n} \sum_{i=1}^{n} \frac{\sum_{j=1}^{o} (y_i^j - y'^j_i)^2}{o} \tag{6}$$

Where n is the number of collected agents, o is the number of objectives, y_i^j and y'^j_i are the actual and predicted normalized objective value j of agent i. The objective values are normalized into the range of 0 and 1 by dividing with the possible maximum objective value.

Then, the number of sub-generation (N) is determined using unit cycle functions, i.e.:

$$N = N_{\max} \times (1 - \sqrt{1 - E^2}). \tag{7}$$

$N_{\max}$ denotes the maximum number of sub-generation. In each sub-generation, offspring are produced using crossover and mutation using the same method of producing offspring at the end of data collection cycle (step E). However, instead of propagate them back to sensor network, they are forwarded to artificial neural network to approximate their objectives value (step F). Then, they are placed in the population pool to be evaluated further together with elites from archive using non-domination ranking. The detail of fitness evaluation is discussed in Section 3.6.

Empty the archive
while true
      Collect the agents that have arrived at base stations. (A)
      Train neural network using the collected agents (B)
      Add collected agents to the population pool. (C)
      Move agents from the archive to the population pool.
      Empty the archive
      for each agent in the population pool
        do **if** Not dominated by all other agents in the population pool. (D)
do   **then** Add the agent to the archive.
      Select elite agents from the archive. (E.)
      Perform genetic operations (crossover/mutation) on elite agents
      if Accuracy of neural network allows to perform approximation
        then Approximate the objective value of offspring using
      neural network then add them to population pool. (F.)
      if Time to broadcast base station pheromone
        then Propagate the offspring to individual nodes in the
      network together with base station pheromone. (G).

Fig. 4. Evolutionary Operations in the El Niño Server

Figure 4 shows the pseudocode of the evolution operations in the El Niño server.

3.6 Fitness Approximation Using Artificial Neural Network

The El Niño server utilizes an artificial neural network to approximate the fitness of each offspring. Artificial neural network is an interconnected network of artificial neurons (or just neurons), a simple processing unit, which can be used for pattern matching or classification. In this work, artificial neural network is used to approximate agent's objective values from agent's gene. The artificial neural network used in this work is called multilayer perceptron, which consists of three layers of neurons; input layer, hidden layer and output layer. The neurons in input layer are fully connected to all neurons in hidden layer and also neurons in hidden layer are fully connected to neurons in output layer. The interconnection between neural has associated weight value ($w_{i,j}$) which can be adjusted in the training process. When an agent's gene is provided to the neurons in input layer as input value, the input value is multiplied with weight before transmitted to a neural in the hidden layer. The neural sums up all of its input and pass the summation to an activation function, usually a sigmoid function. Equation 8 shows this calculation.

$$y_j(x) = K(\sum_i w_{i,j} x_i) \tag{8}$$

From the equation, x_i is the input value i, w_i, j is the weight between neuron i and j, and $K()$ is an activation function. In this work, a sigmoid function which defined by following formula is used:

$$K(x) = \frac{1}{1 + e^{-x}} \tag{9}$$

The output of the activation function ($y(x)$) is send to the connected neurons in output layer. The neurons in output layer performs the same operation and compute the output. Finally, the output of the neural network is the predicted objective values based on the agent's gene at the input layer.

Training through Backpropagation

Before an artificial neural network can be used, it need to be trained. Artificial neural network are trained using training data set, i.e., in this paper, a pair of agent's gene ($\vec{x}$) and agent's objective values ($\vec{y}$). The training process used in this paper is called Backpropagation and it tries to reduce the error between predicted objective values ($\vec{y'}$) and actual objective values ($\vec{y}$) by adjusting the interconnection weight. Following steps are the Backpropagation training process:

1. The predicted output values $\vec{y'}$, i.e., predicted objective values, are computed using the input values $\vec{x}$, i.e., agent's gene, using the equation 8 and 9
2. For each neural k in the output layer, compute the output error δ_k, using:

$$\delta_k = y'(1 - y')(y - y') \tag{10}$$

3. For each neural m in the hidden layer, compute the hidden error δ_m, using:

$$\delta_m = h_m(1 - h_m) \sum_k w_{m,k} \delta_k \tag{11}$$

Where h_m is the output from the equation 8 for neuron m, $w_{m,k}$ is the weight from the hidden neuron m to the output neuron k

4. Then, each weight $w_{i,m}$ between input and hidden layer is updated as follows:

$$w_{i,m} = w_{i,m} + \eta \delta_m x_i \tag{12}$$

Where x_i is the input value that the weight applied to, δ_m is the error associated with the weight and η is the learning rate, usually a positive value less than one and close to zero.

5. Each weight $w_{m,k}$ between hidden and output layer is updated as follows:

$$w_{m,k} = w_{m,k} + \eta \delta_k h_m \tag{13}$$

In this work, a multilayer perceptron with twelve neurons at input layer, thirteen neurons at hidden layer and six neurons at output layer. The number of neurons at input and output layer are corresponding to the gene size and the number of objective values.

4   Simulation Evaluation

This section evaluates El Niño through simulations in terms of its self-configuration, self-optimization and self-healing properties. The El Niño server and runtime are implemented in Java and nesC, respectively. The current codebase of El Niño contains 2,235 lines of Java code and 1,235 lines of nesC code[3]. The El Niño runtime's memory footprint is 3.5 KB in RAM and 36 KB in ROM on a MICA2 node, which equips 4 KB in RAM and 128 KB in ROM. All simulations were carried out with the TOSSIM simulator [7].

A simulated WSN consists of 100 nodes deployed uniformly in a 300x300 meter observation area. A base station is deployed at the area's northwestern corner and linked to the El Niño server via emulated serial port connection. Each node's communication range is 30 meters. Each network link follows the packet loss rate of 5%. The initial sleep period is one minute, and its minimum and maximum are one and five minutes, respectively. The maximum mutation rate is 0.25 (m_{max} in Equation 5). A data prediction model is configured as ARIMA(4, 1, 1), which uses four prior data, one random term and first-order differential. Accelerated evolution uses the learning rate of 0.2 (η in Equations 12 and 13). The maximum number of sub-generations is 15 (N_{max} in Equation 7).

4.1   Simulation Results without Accelerated Evolution

Figure 5 shows the average objective values that agents yield when accelerated evolution is disabled. Each simulation tick corresponds to a duty cycle. Figure 5(a) shows a result in a static network whose conditions never change dynamically. Each objective value converges around the 55th tick. This result shows that, through evolution, agents successfully self-configure their operational parameters and self-optimize their performance by seeking the tradeoffs among conflicting objectives.

Figures 5(b) to 5(f) show how agents perform in dynamic networks. Upon each dynamic change that occurs at the 80th tick, objective values degrade. In Figure 5(b), when 25 nodes are added at random locations, objective values degrade because agents initially have random operational parameters on new nodes. Those agents cannot migrate efficiently to the base station. Also, enough pheromones are not available on new nodes when they are deployed; agents cannot make proper migration decisions on the nodes. In Figure 5(c), randomly-selected 25 nodes fail. As a result, objective values degrade because some agents try to migrate to failed nodes. Figure 5(d), objective values degrade when approximately 25 nodes fail based on a probability distribution. The failure probability of a node follows a Gaussian distribution in which the node at the center of the observation area has the highest failure probability. (This means that the nodes at the middle of the observation area are more likely to fail while the nodes near the area's edges are less likely to fail.) In Figure 5(e), 25% of links between nodes fail. Each link failure is uni-directional. In Figure 5(f), packet loss rate increases from 5% to 25%.

Once objective values degrade due to a dynamic change in the network, agents gradually improve and converge their objective values again. Objective values are mostly

---

[3] http://code.google.com/p/tinydds/

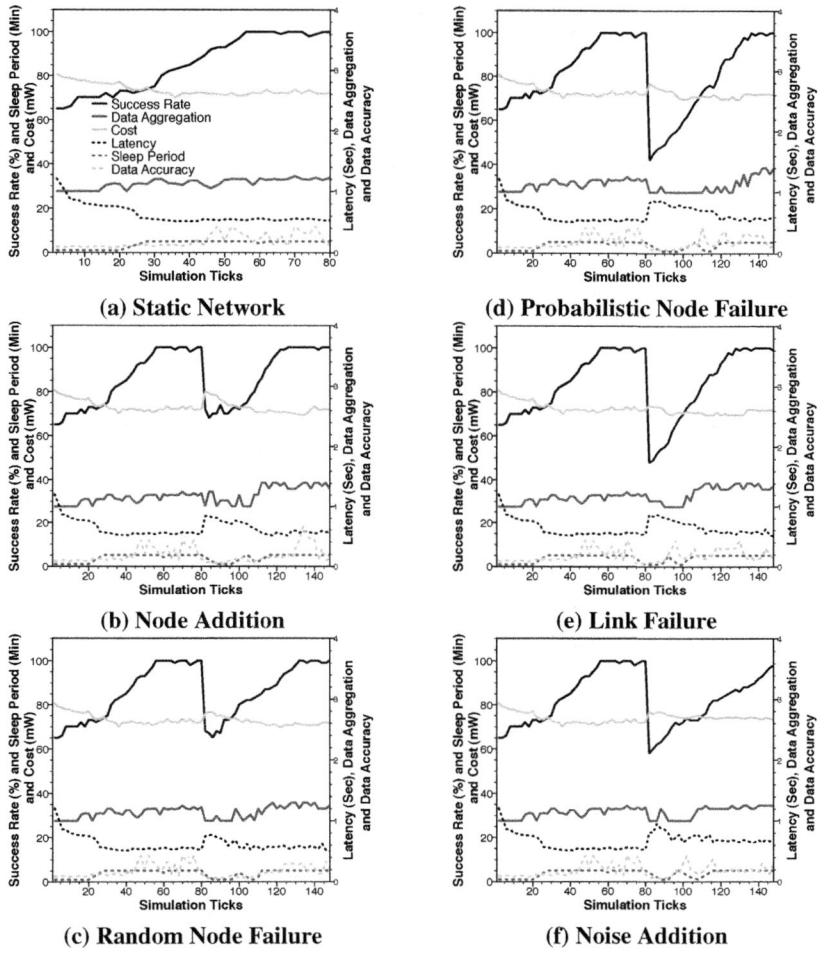

Fig. 5. Performance with Accelerated Evolution Disabled

same before and after each dynamic change. In Figure 5(b), agents yield a higher degree of data aggregation after a node addition because there are more agents migrating in the network and there are higher chances for them to aggregate. Figures 5(b) to 5(f) show that El Niño allows agents to attain a self-healing property; agents autonomously detect and recover from dynamic changes/disruptions in the network. Despite those changes, agents self-configure their operational parameters and self-optimize their performance by evolving their operational parameters.

4.2 Simulation Results with Accelerated Evolution

Figure 6 shows the average objective values that agents yield with accelerated evolution enabled. All simulation configurations are same as the ones used for Figure 5. Figure 6(a) shows that agents yield qualitatively similar performance to the one in

Figure 5(a). However, in Figure 6(a), agents improve all objective values faster than in Figure 5(a). Table 1 compares the results in Figures 5(a) and 6(a), and analyzes the impacts of accelerated evolution on performance convergence in each objective. It shows the number of ticks required for agents to reach particular fractions of the best objective value that they yield with accelerated evolution disabled. For example, the best success rate is 100% with accelerated evolution disabled. Thus, Table 1 shows the number of ticks required for agents to reach the success rate of 90%, 95% and 98% with and without accelerated evolution. It takes only 22 ticks for agents to yield the success rate of 90% with accelerated evolution, while it takes 46 ticks without it. This means that accelerated evolution allows agents to improve their success rate 52.1% faster. (See the column named "Gain by AE" in Table 1.) With respect to the other objectives, agents also gain significant improvement (up to 58.5%) in performance convergence. Figure 6(a) and Table 1 demonstrate that accelerated evolution allows agents to successfully expedite their evolution and augment their self-configuration and self-optimization abilities.

Table 1. Performance with and without Accelerated Evolution (AE) in a Static Network

Objective	Performance Level	Without AE (ticks)	With AE (ticks)	Gain by AE (%)
	90 %	46	22	52.1
Success Rate	95 %	52	34	34.6
	98 %	55	40	27.3
	90 %	29	21	27.5
Data Aggregation	95 %	43	22	48.8
	98 %	70	29	58.5
	90 %	21	20	4.7
Cost	95 %	25	23	8.0
	98 %	67	52	22.3
	90 %	28	24	14.2
Latency	95 %	46	32	30.4
	98 %	56	54	3.57
	90 %	27	27	0
Sleep Period	95 %	28	27	3.5
	98 %	28	28	0
	90 %	68	47	30.9
Data Accuracy	95 %	69	48	30.4
	98 %	69	50	27.5

In Figures 6(b) to 6(f), agents perform similarly to Figures 5(b) to 5(f) in that they exhibit a self-healing property under dynamic changes/disruptions in the network. Upon each change/disruption in the network, agents recover and converge their objective values again. Compared with Figures 5(b) to 5(f), Figures 6(b) to 6(f) illustrate that agents recover their objective values faster with accelerated evolution enabled. Similar to Table 1, Table 2 compares the results in Figures 5(c) and 6(c), and analyzes the impacts of accelerated evolution on performance recovery/convergence in each objective. Accelerated evolution allows agents to consistently gain up to 43.3% improvement in performance recovery/convergence upon random node failures. (See the

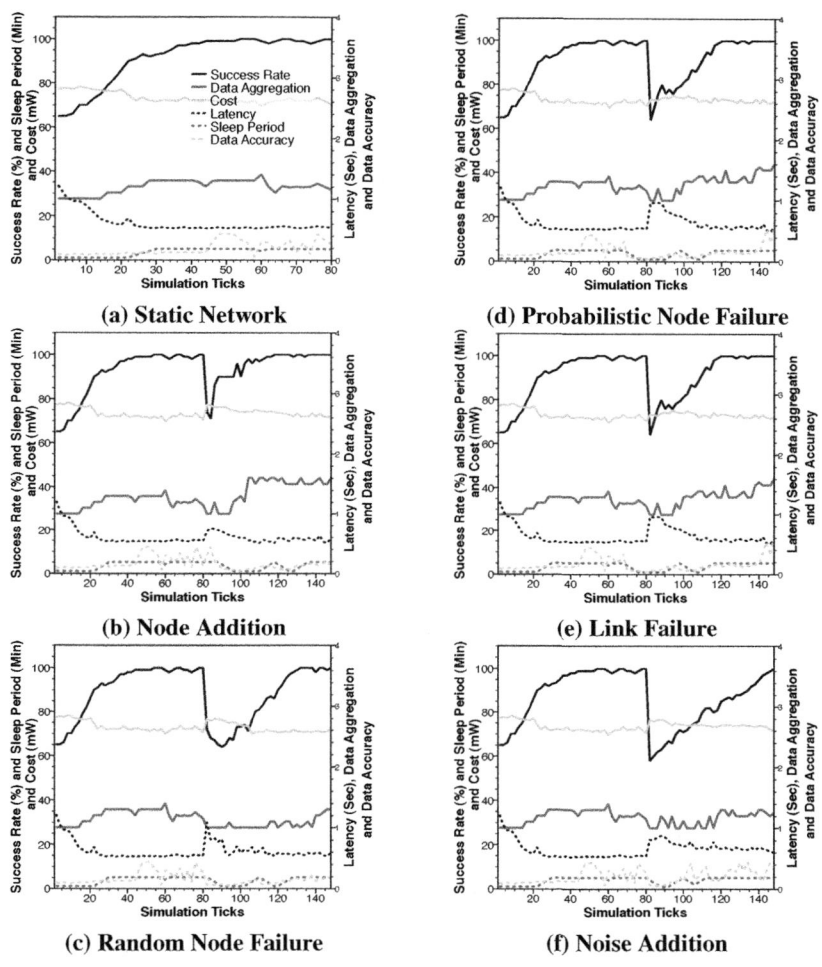

Fig. 6. Performance with Accelerated Evolution Enabled

column named "Gain by AE" in Table 2.) Figures 6(b) to 6(f) and Table 2 demonstrate that accelerated evolution allows agents to successfully expedite their evolution under dynamic changes/disruptions in the network and augment their self-configuration, self-optimization and self-healing abilities.

4.3 Power Efficiency and Performance Approximation Accuracy in Accelerated Evolution

Figure 7 shows the average power consumption on each node as well as the performance approximation accuracy in accelerated evolution. All simulation configurations are same as the ones used for Figure 6. Figure 7(a) shows that power consumption rapidly decreases at the beginning of a simulation and then keeps decreasing gradually over time. In Figures 7(b) to 7(f), power consumption spikes at the 80th simulation tick

Table 2. Performance with and without Accelerated Evolution (AE) in a Dynamic Network in which Nodes Fail Randomly

Objective	Performance Level	Without AE (ticks)	With AE (ticks)	Gain by AE (%)
	90 %	42	40	4.7
Success Rate	95 %	46	44	4.3
	98 %	50	48	4.0
	90 %	68	64	5.8
Data Aggregation	95 %	118	108	8.4
	98 %	119	110	7.5
	90 %	10	10	0
Cost	95 %	28	24	14.2
	98 %	90	74	17.7
	90 %	60	34	43.3
Latency	95 %	90	84	8.6
	98 %	114	112	1.7
	90 %	26	23	11.5
Sleep Period	95 %	26	24	7.7
	98 %	26	24	7.7
	90 %	52	47	9.6
Data Accuracy	95 %	53	48	9.6
	98 %	53	50	5.8

due to dynamic changes/disruptions in the network. However, El Niño allows agents to evolve their operational parameters to new network conditions and reduce power consumption again. Power consumption is mostly same before and after each dynamic change/disruption. Figure 7 demonstrates that El Niño allows agents to be power efficient in dynamic networks as well as static networks through their self-configuration, self-optimization and self-healing abilities.

Figure 7(a) illustrates that accelerated evolution's performance approximation accuracy starts with approximately 55% at the beginning of a simulation and exceeds 75% around the 25th tick. As shown in Figure 6, a higher approximation accuracy contributes to a faster convergence in objective values. In Figures 7(b) to 7(f), approximation accuracy drops at the 80th simulation tick due to dynamic changes/disruptions in the network. However, El Niño learns new network conditions and recovers its approximation accuracy again. Figure 7 shows that El Niño successfully expedites agent evolution (i.e., performance convergence) by improving and recovering its approximation accuracy in both static and dynamic networks.

4.4   Comparison with an Existing Fault-Tolerant Routing Protocol

This section compares El Niño with an existing fault-tolerant adaptive routing protocol, called Wisden [8]. Wisden implements three key mechanisms: low-overhead data time-stamping, wavelet-based data compression and reliable data transport using end-to-end and hop-by-hop recovery.

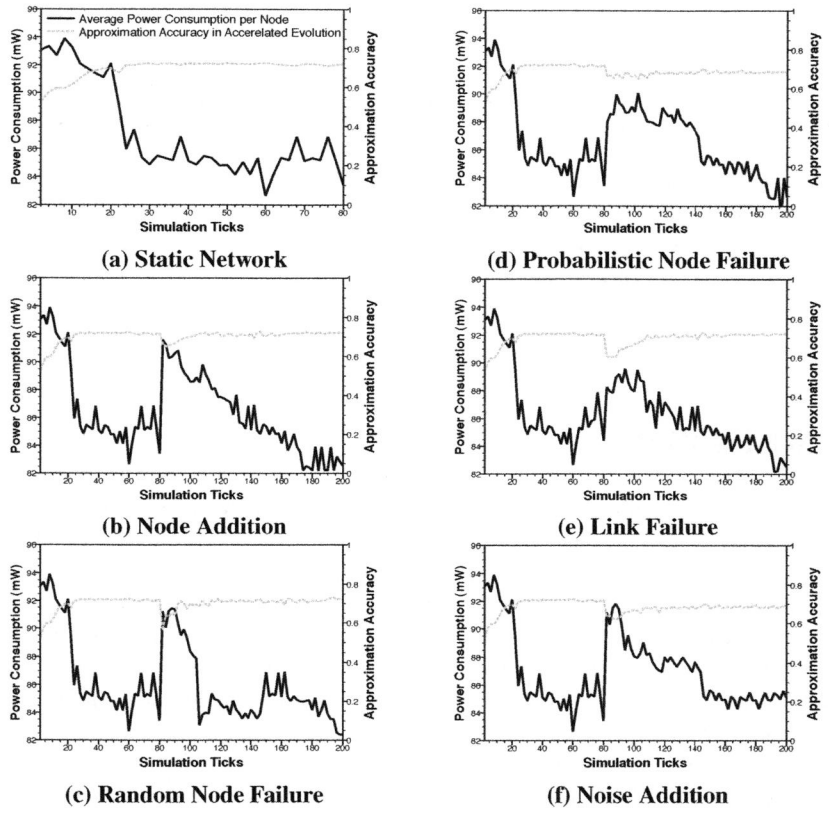

(a) Static Network

(d) Probabilistic Node Failure

(b) Node Addition

(e) Link Failure

(c) Random Node Failure

(f) Noise Addition

Fig. 7. Power Efficiency and Performance Approximation Accuracy in Accelerated Evolution

Figure 8 shows the average objective values that Wisden yields with the simulation configurations used for Figure 6 (c). Only three types of objective values (success rate, latency and cost) are measured because Wisden does not consider tuning data aggregation and sleep period. As Figure 8 illustrates, Wisden improves success rate over time. However, it fails to achieve the success rate of 100% even in between the 0th tick and the 80th tick due to constant packet losses. (Note that the packet loss rate of 5% is assumed throughout a simulation.) Wisden does not sufficiently improve latency and cost as El Niño does (Figure 6 (c)). Moreover, when nodes fail at the 80th tick, latency spikes to approximately 2.5 seconds because Wisden performs end-to-end recovery to obtain missing data. Latency stays high until the 120th tick and decreases to approximately 1 second around the 140th tick. Figures 6 (c) and 8 demonstrate that El Niño outperforms Wisden in that El Niño improves its performance with respect to all of six objectives and recovers the performance more quickly than Wisden under dynamic changes/disruptions in the network.

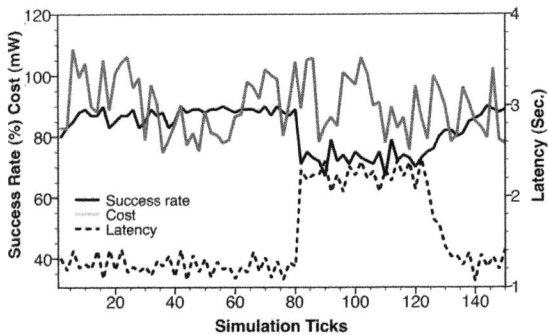

Fig. 8. Performance of Wisden

5   Related Work

This paper describes a set of extensions to the authors' previous work [9]. The previous work studied an evolutionary adaptation mechanism for WSNs, called MONSOON. El Niño extends MONSOON by investigating accelerated evolution and considering more operational parameters. El Niño considers twelve operational parameters that often conflict with each other, while MONSOON considers four parameters that do not conflict. El Niño is designed to solve harder (or more realistic) optimization problems in WSNs. (As discussed in Section 1, multiobjective optimization becomes harder to solve when parameters conflict.)

Baldi et al. propose a cost (or fitness) function that comprises conflicting performance objectives in UWB wireless networks [10]. They include data transmission cost, power consumption, latency, node/link reliability and link interference. These objectives are similar to the ones El Niño considers. However, in [10], the total cost (or fitness) is computed as a weighted sum of objective values. This means that application designers need to manually configure every weight value in a fitness function through labor-intensive trial and errors. In El Niño, no manually-configured parameters exist for elite selection because of a domination ranking mechanism. El Niño minimizes configuration costs for application designers. Moreover, El Niño does not require each node to have global network information as [10] does.

Genetic algorithms (GAs) have been investigated in various aspects in WSNs; for example, routing [11, 12, 13, 14, 15], data processing [16], localization [17, 18], node placement [19,20] and object tracking [15]. None of them study a mechanism equivalent to accelerated evolution in El Niño. Also, all of them use fitness functions, each of which combines multiple objective values as a weighted sum to rank agents/genes in elite selection. As discussed above, it is always non-trivial to manually configure weight values in a fitness function. In contrast, El Niño eliminates parameters in elite selection by design. Moreover, network dynamics and disruptions are not considered in [11, 12, 14, 17, 19, 20, 15] do not assume dynamic WSNs.

Beyond these classical GAs, multiobjective GAs (MOGAs) have also been investigated in WSNs; for example, for routing [21, 22, 23, 24], node placement [25, 26, 27, 28, 29] and duty cycle management [30]. Unlike El Niño, all of these work assume static WSNs, not dynamic WSNs. They do not investigate a mechanism equivalent to accelerated evolution in El Niño. In [21, 22, 23], MOGAs are studied to optimize migration routes (or itineraries) that mobile agents follow from a base station to cluster head nodes. On the contrary, in El Niño, agents make their migration and other behavior decisions by themselves. El Niño optimizes agents' operational parameters, not their migration routes.

Sin et al. follow the agent design proposed in El Niño and exploit MOGAs to optimize the operational parameters of agents. However, they do not consider dynamic WSNs as El Niño does. Moreover, El Niño studies accelerated evolution, which is beyond the scope of [24].

Liu et al. and Li et al. use ARIMA to predict sensor reading and reduce power consumption [31, 32]. Each node compares predicted and actual sensor readings and transmits sensor data to a base station only when a prediction error exceeds a certain threshold. This way, nodes can reduce power consumption by eliminating unnecessary data transmissions. Each base station performs the same prediction method when it does not receive sensor data from a node, and substitutes the missing data with predicted data. El Niño takes a different approach for power efficiency. It uses ARIMA to compute data accuracy and aims to reduce power consumption by tuning three operational parameters (sleep period, sensing rate and reporting rate) with respect to two conflicting objectives: sleep period and data accuracy (Sections 3.1 and 3.2). It does not have to configure and use a threshold for prediction/accuracy error. Moreover, El Niño assumes dynamic networks and train its ARIMA model online, while Liu et al. assume offline training in static networks.

Vassev et al. propose a formal method to specify and generate prototype operation models for sensor nodes [33, 34]. Generated models allow nodes to operate in a self-managing manner. A model is constructed with a set of adaptation policies, each of which pairs an event and an action(s) that a node performs. It is assumed that policies are supplied in advance. This means that application designers need to anticipate a variety of events and pre-define corresponding actions. El Niño takes a different approach for self-managing WSNs. It never assumes events in advance. Behavior policies are not predefined and fixed. El Niño autonomously seeks the optimal behavior policies under dynamic network conditions through evolutionary process augmented by a notion of accelerated evolution.

6   Conclusion

This paper proposes and evaluates an autonomic sensor networking framework, called El Niño, for data collection applications in WSNs. El Niño allows agents to evolve and adapt their operational parameters to dynamic network conditions in a self-configuring, self-optimizing and self-healing manner by seeking the optimal tradeoffs among conflicting performance objectives. El Niño's accelerated evolution allows agents to expedite their evolution and efficiently adapts to network dynamics and disruptions.

References

1. Karl, H., Willig, A.: Protocols and Architectures for Wireless Sensor Networks. Wiley Interscience (2007)
2. Han, Q., Hakkarinen, D., Boonma, P., Suzuki, J.: Quality-Aware Sensor Data Collection. International Journal of Sensor Networks 7, 127–140 (2010)
3. Gershenson, C., Heylighen, F.: When Can We Call a System Self-Organizing? In: Banzhaf, W., Ziegler, J., Christaller, T., Dittrich, P., Kim, J.T. (eds.) ECAL 2003. LNCS (LNAI), vol. 2801, pp. 606–614. Springer, Heidelberg (2003)
4. Seeley, T.: The Wisdom of the Hive. Harvard University Press (2005)
5. Levis, P., Madden, S., Polastre, J., Szewczyk, R., Whitehouse, K., Woo, A., Gay, D., Hill, J., Welsh, M., Brewer, E., et al.: TinyOS: An operating system for sensor networks. In: Ambient Intelligence, pp. 115–148. Springer, Heidelberg (2005)
6. Shnayder, V., Hempstead, M., Chen, B.-R., Werner-Allen, G., Welsh, M.: Simulating the power consumption of large-scale sensor network applications. In: Proc. of IEEE Conference on Embedded Networked Sensor Systems (2004)
7. Levis, P., Lee, N., Welsh, M., Culler, D.: TOSSIM: Accurate and Scalable Simulation of Entire TinyOS Applications. In: Proc. of IEEE Int'l Conference on Embedded Networked Sensor System, pp. 126–137 (2003)
8. Xu, N., Rangwala, S., Chintalapudi, K.K., Ganesan, D., Broad, A., Govindan, R., Estrin, D.: Wireless Sensor Network for Structural Monitoring. In: Proc. of ACM Int'l Conference on Embedded Networked Sensor Systems, pp. 13–24 (2005)
9. Boonma, P., Suzuki, J.: Exploring Self-star Properties in Cognitive Sensor Networking. In: Proc. of IEEE/SCS Int'l Symposium on Performance Evaluation of Computer and Telecommunication Systems, pp. 36–43 (2008)
10. Baldi, P., Nardis, L.D., Benedetto, M.G.D.: Modeling and Optimization of UWB Communication Networks Through a Flexible Cost Function. IEEE J. on Sel. Areas in Commun. 20, 1733–1744 (2002)
11. Khanna, R., Liu, H., Chen, H.: Self-organisation of Sensor Networks using Genetic Algorithms. Inderscience Int'l J. of Sensor Networks 1(3), 241–252 (2006)
12. Hussain, S., Matin, A.W., Islam, O.: Genetic Algorithm for Hierarchical Wireless Sensor Networks. Journal of Networks 2(5), 87–97 (2007)
13. Jin, S., Zhou, M., Wu, A.S.: Sensor Network Optimization using a Genetic Algorithm. In: Proc. of IIIS World Multi-Conference on Systemics, Cybernetics and Informatics (2003)
14. Ferentinos, K.P., Tsiligiridis, T.A.: Adaptive Design Optimization of Wireless Sensor Networks using Genetic Algorithms. Computer Networks: The International Journal of Computer and Telecommunications Networking 51(4), 1031–1051 (2007)
15. Buczaka, A.L., Wangb, H.: Optimization of Fitness Functions with Non-ordered Parameters by Genetic Algorithms. In: Proc. of IEEE Congress on Evolutionary Computation, pp. 199–206 (2001)
16. Hauser, J., Purdy, C.: Sensor Data Processing using Genetic Algorithms. In: Proc. of IEEE Midwest Symposium on Circuits and Systems, pp. 1112–1115 (2000)
17. Tam, V., Cheng, K.Y., Lui, K.S.: Using Micro-Genetic Algorithms to Improve Localization in Wireless Sensor Networks. Journal of Communications 1(4), 1–10
18. Zhang, Q., Wang, J., Jin, C., Ye, J., Ma, C., Zhang, W.: Genetic Algorithm Based Wireless Sensor Network Localization. In: Proc. of IEEE Int'l Conference on Natural Computation, pp. 608–613 (2008)
19. Guo, H.Y., Zhang, L., Zhang, L.L., Zhou, J.X.: Optimal Placement of Sensors for Structural Health Monitoring using Improved Genetic Algorithms. Smart Materials and Structures 13(3), 528–534 (2004)

20. Zhao, J., Wen, Y., Shang, R., Wang, G.: Optimizing Sensor Node Distribution with Genetic Algorithm in Wireless Sensor Network. In: Yin, F.-L., Wang, J., Guo, C. (eds.) ISNN 2004. LNCS, vol. 3174, pp. 242–247. Springer, Heidelberg (2004)
21. Rajagopalan, R., Mohan, C., Varshney, P., Mehrotra, K.: Multi-objective Mobile Agent Routing in Wireless Sensor Networks. In: Proc. of IEEE Congress on Evolutionary Computation, pp. 1730–1737 (2005)
22. Rajagopalan, R., Varshney, P.K., Mehrotra, K.G., Mohan, C.K.: Fault Tolerant Mobile Agent Routing in Sensor Networks: A Multi-Objective Optimization Approach. In: Proc. of IEEE Upstate New York Workshop on Communication and Networking (2005)
23. Xue, F., Sanderson, A., Graves, R.: Multi-Objective Routing in Wireless Sensor Networks with a Differential Evolution Algorithm. In: Proc. of IEEE Int'l Conference on Networking, Sensing and Control, pp. 880–885 (2006)
24. Sin, H., Lee, J., Lee, S., Yoo, S., Lee, S., Lee, J., Lee, Y., Kim, S.: Agent-based Framework for Energy Efficiency in Wireless Sensor Networks. World Academy of Science, Engineering and Technology 46, 305–309 (2008)
25. Jourdan, D.B., de Weck, O.L.: Multi-Objective Genetic Algorithm for the Automated Planning of a Wireless Sensor Network to Monitor A Critical Facility. In: Proc. of SPIE Defense and Security Symposium, pp. 565–575 (2004)
26. Rajagopalan, R., Varshney, P.K., Mohan, C.K., Mehrotra, K.G.: Sensor Placement for Energy Efficient Target Detection in Wireless Sensor Networks: A multi-objective Optimization Approach. In: Proc. of Annual Conference on Information Sciences and Systems (2005)
27. Raich, A.M., Liszkai, T.R.: Multi-Objective Genetic Algorithm Methodology for Optimizing Sensor Layouts to Enhance Structural Damage Identification. In: Proc. of Int'l Workshop on Structural Health Monitoring, pp. 650–657 (2003)
28. Jia, J., Chen, J., Chang, G., Tan, Z.: Energy Efficient Coverage Control in Wireless Sensor Networks based on Multi-Objective Genetic Algorithm. Computers & Mathematics with Applications 57(11-12), 1756–1766 (2009)
29. Molina, G., Alba, E., Talbi, E.G.: Optimal Sensor Network Layout Using Multi-Objective Metaheuristics. J. of Universal Computer Science 14(15), 2549–2565 (2008)
30. Yang, E., Erdogan, A.T., Arslan, T., Barton, N.: Multi-Objective Evolutionary Optimizations of a Space-Based Reconfigurable Sensor Network under Hard Constraints. Soft Computing - A Fusion of Foundations, Methodologies and Applications 15(1), 25–36 (2011)
31. Liu, C., Wu, K., Tsao, M.: Energy Efficient Information Collection with the ARIMA model in Wireless Sensor Networks. In: Proc. of IEEE Global Telecommunication Conference, pp. 2470–2474 (2005)
32. Li, M., Ganesan, D., Shenoy, P.: PRESTO: Feedback-Driven Data Management in Sensor Networks. In: Proc. of ACM/USENIX Symposium on Networked Systems Design and Implementation, pp. 311–324 (2006)
33. Vassev, E., Hinchey, M., Nixon, P.: Prototyping Home Automation Wireless Sensor Networks with ASSL. In: Proc. of ACM Int'l Conference on Autonomic Computing, pp. 71–72 (2010)
34. Vassev, E., Hinchey, M., Nixon, P.: Developing Intelligent Sensor Networks: A Technological Convergence Approach. In: Proc. of ACM/IEEE Int'l Workshop on Software Engineering for Sensor Network Applications, pp. 66–71 (2010)

Developing Autonomic Properties
for Distributed Pattern-Recognition Systems
with ASSL*
A Distributed MARF Case Study

Emil Vassev[1] and Serguei A. Mokhov[2]

[1] Lero - The Irish Software Engineering Research Centre,
University of Limerick, Limerick, Ireland
emil@vassev.com
[2] Faculty of Engineering and Computer Science, Concordia University,
1455 de Maisonneuve Blvd. W, Montreal, QC, Canada
mokhov@cse.concordia.ca

Abstract. We discuss our research towards developing special properties that introduce autonomic behavior in distributed pattern-recognition systems. In our approach we use ASSL (Autonomic System Specification Language) to formally develop such properties for DMARF (Distributed Modular Audio Recognition Framework). These properties enhance DMARF with an autonomic middleware that manages the four stages of the framework's pattern-recognition pipeline. DMARF is a biologically inspired system employing pattern recognition, signal processing, and natural language processing helping us process audio, textual, or imagery data needed by a variety of scientific applications, e.g., biometric applications. In that context, the notion go autonomic DMARF (ADMARF) can be employed by autonomous and robotic systems that theoretically require less-to-none human intervention other than data collection for pattern analysis and observing the results. In this article, we explain the ASSL specification models for the autonomic properties of DMARF.

Keywords: autonomic computing, formal methods, ASSL, DMARF.

1 Introduction

Today, we face the challenge of hardware and software complexity that appears to be the biggest threat to the continuous progress in IT. Many initiatives towards complexity reduction in both software and hardware have arisen with the

---

* This work was supported in part by an IRCSET postdoctoral fellowship grant (now termed as EMPOWER) at University College Dublin, Ireland, by the Science Foundation Ireland grant 03/CE2/I303_1 to Lero (the Irish Software Engineering Research Centre), and by the Faculty of Engineering and Computer Science of Concordia University, Montreal, Canada.

M.L. Gavrilova et al. (Eds.): Trans. on Comput. Sci. XV, LNCS 7050, pp. 130–157, 2012.
© Springer-Verlag Berlin Heidelberg 2012

advent of new theories and paradigms. Autonomic computing (AC) [13] promises reduction of the workload needed to maintain complex systems by transforming them into self-managing autonomic systems. The AC paradigm draws inspiration from the human body's *autonomic nervous system* [3]. The idea is that software systems can manage themselves and deal with dynamic requirements, as well as unanticipated threats, automatically, just as the body does, by handling complexity through self-management.

Pattern recognition is a widely used biologically inspired technique in the modern computer science. Algorithms for image and voice recognition have been derived from the human brain, which uses pattern recognition to recognize shapes, images, voices, sounds, etc. In this research, we applied the principles of AC to solve specific problems in distributed pattern-recognition systems, such as availability, security, performance, etc. where the health of a distributed pipeline is important. We tackled these issues by introducing self-management into the system behavior. As a proof-of-concept (PoC) case study, we used ASSL [18,25] to develop the autonomic self-management properties for DMARF [10], which is an intrinsically complex pipelined distributed system composed of multi-level operational layers. The ASSL framework helped us develop the self-managing features first, which we then integrated into DMARF. In this paper, based on our results, we provide an attempt to generalize our experience on any similar-scale distributed pipelined pattern-recognition system.

1.1 Problem Statement and Proposed Solution

Distributed MARF (DMARF) could not be used in autonomous systems of any kind as-is due to lack of provision for such a use by applications that necessitate the self-management requirements. Extending DMARF directly to support the said requirements is a major redesign and development effort to undertake for an open-source project. Moreover, such and extended autonomic DMARF must be validated and tested, since there is no immediate guarantee that the properties the latter has been augmented with are intrinsically correct.

In our approach, we provide the methodology for the initial proof-of-concept. We specify with ASSL a number of autonomic properties for DMARF, such as self-healing [24], self-optimization [23], and self-protection [12]. The implementation of those properties is generated automatically by the ASSL framework in the form of a special wrapper Java code that provides an autonomic layer implementing the DMARF's autonomic properties. In addition, the latter are formally validated with the ASSL's mechanisms for consistency and model checking. Model checking is performed on the generated Java code, where the ASSL relies on the Java PathFinder [2] tool developed by NASA Ames.

The rest of this article is organized as follows. In Section 2, we briefly review the field of autonomic computing and describe both ASSL and DMARF frameworks. Section 3 presents details of the ASSL specification models for autonomic properties of DMARF. Finally, Section 4 presents some concluding remarks and future work.

2   Background

The vision and metaphor of AC [13] is to apply the principles of self-regulation and complexity hiding. The AC paradigm emphasizes reduction of the workload needed to maintain complex systems by transforming those into self-managing autonomic systems (AS). The idea is that software systems shall automatically manage themselves just as the human body does. Nowadays, a great deal of research effort is devoted to developing AC tools. Such a tool is the ASSL framework, which helps AC developers with problem specification, system design, system analysis and evaluation, and system implementation.

ASSL was initially developed by Vassev at Concordia University, Montreal, Canada [25] and since then it has been successfully applied to the development of a variety of autonomic systems including distributed ones. For example, ASSL was used to develop autonomic features and generate prototype models for two NASA missions – the ANTS (Autonomous Nano-Technology Swarm) concept mission (where a thousand of picospacecraft work cooperatively to explore the asteroid belt [17]), and the Voyager mission [16]. In both cases, there have been developed autonomic prototypes to simulate the autonomic properties of the space exploration missions and validate those properties through the simulated experimental results. The targeted autonomic properties were: self-configuring [19], self-healing [20], and self-scheduling [22] for ANTS, and autonomic image processing for Voyager [21]. In general, the development of these properties required a two-level approach, i.e., they were specified at the individual spacecraft level and at the level of the entire system. Because ANTS is intrinsically distributed system composed of many autonomous spacecraft, that case study required individual specification of the autonomic properties of each individual spacecraft member of the ANTS swarm.

2.1   ASSL

The Autonomic System Specification Language (ASSL) [18,25] approaches the problem of formal specification and code generation of autonomic systems (ASs) within a framework. The core of this framework is a special formal notation and a toolset including tools that allow ASSL specifications be edited and validated. The current validation approach in ASSL is a form of consistency checking (handles syntax and consistency errors) performed against a set of semantic definitions. The latter form a theory that aids in the construction of correct AS specifications. Moreover, from any valid specification, ASSL can generate an operational Java application skeleton.

Overall, ASSL considers autonomic systems (ASs) as composed of autonomic elements (AEs) communicating over interaction protocols. To specify those, ASSL is defined through formalization of tiers. Over these tiers, ASSL provides a multi-tier specification model that is designed to be scalable and exposes a judicious selection and configuration of infrastructure elements and mechanisms needed by an AS. The ASSL tiers and their sub-tiers (cf. Figure 1) are abstractions of different aspects of the AS under consideration. They aid not only to

```
I. Autonomic System (AS)
  * AS Service-level Objectives
  * AS Self-managing Policies
  * AS Architecture
  * AS Actions
  * AS Events
  * AS Metrics
II. AS Interaction Protocol (ASIP)
  * AS Messages
  * AS Communication Channels
  * AS Communication Functions
III. Autonomic Element (AE)
  * AE Service-level Objectives
  * AE Self-managing Policies
  * AE Friends
  * AE Interaction Protocol (AEIP)
    - AE Messages
    - AE Communication Channels
    - AE Communication Functions
    - AE Managed Elements
  * AE Recovery Protocol
  * AE Behavior Models
  * AE Outcomes
  * AE Actions
  * AE Events
  * AE Metrics
```

Fig. 1. ASSL Multi-Tier Model

specification of the system at different levels of abstraction, but also to reduction of the complexity, and thus, to improving the overall perception of the system.

There are three major tiers (three major abstraction perspectives), each composed of sub-tiers (cf. Figure 1):

- *AS tier* – presents a general and global AS perspective, where we define the general autonomic system rules in terms of *service-level objectives (SLO)* and *self-management policies*, *architecture topology* and *global actions*, *events* and *metrics* applied in these rules.
- *AS Interaction Protocol (ASIP) tier* – forms a communication protocol perspective, where we define the means of communication between AEs. An ASIP is composed of *channels*, *communication functions*, and *messages*.
- *AE tier* – forms a unit-level perspective, where we define interacting sets of individual AEs with their own behavior. This tier is composed of AE rules (*SLO* and *self-management policies*), an *AE interaction protocol* (AEIP), *AE friends* (a list of AEs forming a circle of trust), *recovery protocols*, special *behavior models* and *outcomes*, *AE actions*, *AE events*, and *AE metrics*.

The AS Tier specifies an AS in terms of *service-level objectives* (AS SLOs), *self-management policies*, *architecture topology*, *actions*, *events*, and *metrics* (cf. Figure 1). The AS SLOs are a high-level form of behavioral specification that help developers establish system objectives (e.g., performance). The self-management policies could be any of (but not restricted to) the four so-called self-CHOP policies defined by the AC IBM blueprint: *self-configuring, self-healing, self-optimizing* and *self-protecting* [4]. These policies are event-driven and trigger the execution of actions driving an AS in critical situations. The metrics constitute a set of parameters and observables controllable by an AS. At the ASIP Tier, the ASSL framework helps developers specify an AS-level interaction protocol as a public communication interface, expressed with special *communication channels*, *communication functions* and *communication messages*. At the AE Tier, the ASSL formal model exposes specification constructs for the specification of the system's AEs.

Conceptually, AEs are considered to be analogous to software agents able to manage their own behavior and their relationships with other AEs. These relationships are specified at both ASIP and AEIP tiers. Whereas ASIP specifies an AS-level *interaction protocol* that is public and accessible to all the AEs of an AS and to *external systems* communicating with that very AS, the AEIP tier is normally used to specify a *private communication protocol* used by an AE to communicate only with: 1) trusted AEs, i.e., AEs declared as "AE Friends" (cf. Figure 1); and 2) special controlled *managed elements*. Therefore, two AEs exchange messages over an AEIP only if they are *friends*, thus revealing the need for special negotiation messages specified at ASIP to discover new friends at runtime.

Note that ASSL targets only the AC features of a system and helps developers clearly distinguish the AC features from the system-service features. This is possible, because with ASSL we model and generate special AC wrappers in the form of ASs that embed the components of non-AC systems. The latter are considered as *managed elements*, controlled by the AS in question. A managed element can be any software or hardware system (or sub-system) providing services. Managed elements are specified per AE (they form an extra layer at the AEIP cf. Figure 1) where the emphasis is on the control interface. It is important also to mention that the ASSL tiers and sub-tiers are intended to specify different aspects of an AS, but it is not necessary to employ all of them in order to model such a system. For a simple AS we need to specify 1) the AEs providing self-managing behavior intended to control the managed elements associated with an AE; and 2) the communication interface. Here, self-management policies must be specified to provide such self-managing behavior at the level of AS (the AS Tier) and at the level of AE (AE Tier). The self-management behavior of an ASSL-developed AS is specified with the self-management policies. These policies are specified with special ASSL constructs termed *fluents* and *mappings* [18,25]. A fluent is a state where an AS enters with *fluent-activating events* and exits with *fluent-terminating events*. A mapping connects fluents with particular actions to be undertaken. Usually, an ASSL specification is built around self-management policies, which make that specification AC-driven. The policies themselves are driven by events and actions determined deterministically. Figure 2 presents a sample specification of an ASSL self-healing policy.

For more details on the ASSL multi-tier specification model and the ASSL framework toolset, please refer to [18,25].

```
ASSELF_MANAGEMENT {
 SELF_HEALING {
  FLUENT inLosingSpacecraft {
   INITIATED_BY { EVENTS.spaceCraftLost }
   TERMINATED_BY { EVENTS.earthNotified }
    }
  MAPPING {
   CONDITIONS { inLosingSpacecraft }
   DO_ACTIONS { ACTIONS.notifyEarth }
  }
 }
} // ASSELF_MANAGEMENT
```

Fig. 2. Self-management Policy

2.2 Distributed MARF

DMARF [10] is based on the classical MARF whose pipeline stages were made into distributed nodes. The Modular Audio Recognition Framework (MARF) [5] is an open-source research platform and a collection of pattern recognition, signal processing, and natural language processing (NLP) algorithms written in Java and arranged into a modular and extensible framework facilitating addition of new algorithms for use and experiments by scientists. MARF can run distributively over the network, run stand-alone, or may just act as a library in applications. MARF has a number of algorithms implemented for various pattern recognition and some signal processing tasks. The backbone of MARF consists of pipeline stages that communicate with each other to get the data they need in a chained manner.

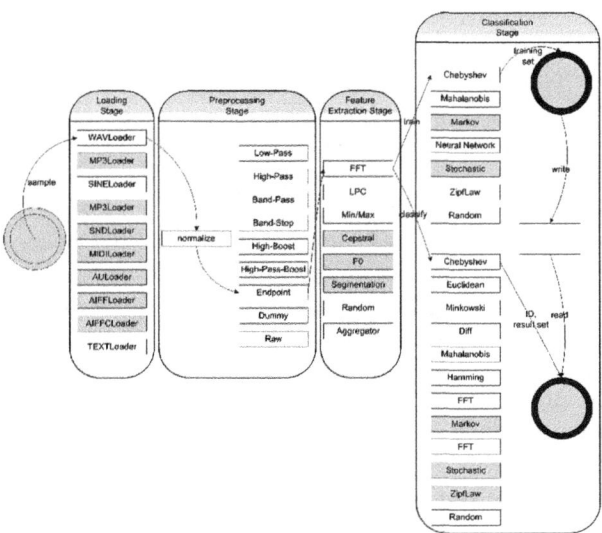

Fig. 3. MARF's Pattern Recognition Pipeline

In general, MARF's pipeline of algorithm implementations is presented in Figure 3 (where the implemented algorithms are grouped in white boxes, and the stubs or in progress algorithms are grouped in gray). The pipeline consists of the four core stages grouping the similar kinds of algorithms: (1) sample loading, (2) preprocessing, (3) feature extraction, and (4) training/classification. MARF's distributed extension, DMARF [10] allows the stages of the pipeline to run as distributed nodes as well as a front-end. The basic stages and the front-end were implemented without backup recovery or hot-swappable capabilities at this point; just communication over Java RMI [26], CORBA [14], and XML-RPC WebServices [15]. There is also an undergoing project on the intensional scripting language, MARFL [7] to allow scripting MARF tasks and applications.

There are various applications that test and employ MARF's functionality and serve as examples of how to use MARF. High-volume processing of recorded audio, textual, or imagery data are possible pattern-recognition and biometric applications of DMARF. In this work, most of the emphasis is on audio processing, such as conference recordings with purpose of attribution of said material to identities of speakers. Another emphasis is on processing a bulk of recorded phone conversations in a police department for forensic analysis [8] and subject identification and classification. See the cited works and references therein for more details on MARF and applications.

3   Making Distributed Pipelined Systems Autonomic with ASSL: Autonomic DMARF Case Study

In general, ASSL helps to design and generate special *autonomic wrappers* in the form of AEs that embed one or more system components. The latter are considered as *managed elements* (cf. Section 2.1) that present one or more *single nodes* of a distributed system. Therefore, for each distributed node, we ideally specify with ASSL a single AE that introduces an autonomic behavior to that node. All the AEs are specified at the *AE Tier* and the global autonomic behavior of the entire system is handled by specifications at the *AS Tier* (cf. Figure 1). As shown in Section 2.1, we rely on a rich set of constructs, such as *actions, events*, and *metrics* to specify special self-management policies driving the nodes of a distributed system in situations requiring autonomic behavior. Moreover, with ASSL, we specify special interaction protocols (ASIP and AEIP) that help those nodes exchange messages and synchronize on common autonomic behavior. In this section we demonstrate how ASSL may be applied to an inherently distributed system such as DMARF. The novelty in our approach is safeguarding the distributed pipeline, which is not possible with plain distributed systems. Therefore, with this case study we not only demonstrate the applicability of ASSL to distributed systems but also validate that ASSL may successfully be applied to *pipelined distributed systems*.

DMARF's capture as an AS primarily covers the autonomic behavior of the distributed pattern-recognition pipeline. We examine properties that apply to DMARF and specify in detail the self-CHOP aspects of it. If we look a the

DMARF pipeline as a whole, we see that there should be at least one instance of every stage somewhere on the network. There are four main core pipeline stages and an application-specific stage that initiates pipeline processing. If one of the core stages goes offline, the pipeline stalls and to recover it has the following options: 1) use of a replacement node; 2) recovery of the failed node; or 3) rerouting the pipeline through a different node with the same service functionality as the failed one.

In order to make DMARF autonomic, we need to add *automicity* (autonomic computing behavior) to the DMARF behavior. We add a special *autonomic manager* (AM) to each DMARF stage. This makes the latter AEs, those composing an autonomic DMARF (ADMARF) capable of self-management.

3.1   Self-Healing

A DMARF-based system should be able to recover itself through replication to keep at least one route of the pipeline available. There are two types of replication: 1) the replication of a service, which essentially means that we increase the number of nodes per core stage (e.g. two different hosts provide preprocessing services as active replication, so if one goes down, the pipeline is still not stalled; if both are up they can contribute to load balancing, which is a part of the self-optimization autonomic property); and 2) replication within the node itself. If all nodes of a core stages go down, the stage preceding it is responsible to start up a temporary one on the host of the preceding stage, set it up to repair the pipeline. This is the hard replication needed to withstand stall faults, where it is more vulnerable and not fault-tolerant. In the second case, denoting passive replication of the same node (or even different nodes) losing a primary or a replica is not as serious as in the first case because such a loss does not produce a pipeline stall and it is easier to self-heal after a passive replica loss. Restart and recovery of the failed node without replicas is another possibility for self-healing for DMARF. Technically, it may be tried prior or after the replica kicks in.

In the course of this project, we used ASSL to specify the self-healing behavior of ADMARF by addressing specific cases related to *node replacement* (service replica) and *node recovery*, shown in Figure 4.

The following sub-sections describe the ASSL specification of the self-healing algorithm revealed here. We specified this algorithm as an ASSL self-healing policy spread on both system (AS tier) and autonomic element (AE tier) levels where *events, actions, metrics*, and special *managed element interface functions* are used to incorporate the self-healing behavior in ADMARF (cf. Appendix A). Note that due to space limitations Appendix A presents a partial ASSL specification where only one AE (DMARF stage) is specified. The full specification specifies all the four DMARF stages.

AS Tier Specification for Self-Healing. At the AS tier we specify the global ADMARF self-healing behavior. To specify the latter, we use an ASSL SELF_HEALING self-management policy (cf. Figure 5). Here we specified a single *fluent* mapped to an *action* via a *mapping*.

1 ADMARF monitors its run-time performance and in case of performance degradation notifies the problematic DMARF stages to start self-healing;

2 Every notified DMARF stage (note that this is an AE) analyzes the problem locally to determine its nature: a node is down or a node is not healthy (does not perform well);

3 **if** *A node is down* **then**

      `// The following node-replacement algorithm is followed by the AM`
          `of the stage:`

4   AM strives to find a replica note of the failed one;

5   **if** *replica found* **then**

6 | next redirect computation to it;

7   **end**

8   **if** *replica not found* **then**

9   report the problem. Note that the algorithm could be extended with a few more steps where the AM contacts the AM of the previous stage to organize pipeline reparation;

10   **end**

11 **end**

12 **if** *A node does not perform well* **then**

      `// The following node-recovery algorithm is followed:`

13   AM starts the recovery protocol for the problematic node;

14   **if** *recovery successful* **then**

15 | do nothing;

16   **end**

17   **if** *recovery unsuccessful* **then**

18 | AM strives to find a replica node of the failed one;

19   **end**

20   **if** *replica found* **then**

21 | next redirect computation to it;

22   **end**

23   **if** *replica not found* **then**

24 | report the problem;

25   **end**

26 **end**

Fig. 4. DMARF Self-Healing Algorithm

```
ASSELF_MANAGEMENT {
  SELF_HEALING {
    // a performance problem has been detected
    FLUENT inLowPerformance {
      INITIATED_BY { EVENTS.lowPerformanceDetected }
      TERMINATED_BY {    EVENTS.performanceNormalized,
                      EVENTS.performanceNormFailed }
    }
    MAPPING {
      CONDITIONS { inLowPerformance }
      DO_ACTIONS { ACTIONS.startSelfHealing }
    }
  }
} // ASSELF_MANAGEMENT
```

Fig. 5. AS Tier SELF_HEALING Policy

```
ASSLO {
  SLO performance {
    FOREACH member in AES {
      member.AESLO.performance
    }
  }
}
....
EVENTS {  // these events are used in the fluents specification
  EVENT lowPerformanceDetected {
    ACTIVATION { DEGRADED { ASSLO.performance } } }
  EVENT performanceNormalized {
    ACTIVATION { NORMALIZED { ASSLO.performance } } }
  EVENT performanceNormFailed {
    ACTIVATION {
      OCCURRED { AES.STAGE_AE.EVENTS.selfHealingFailed } } }
} // EVENTS
```

Fig. 6. AS Tier SLO and Events

Thus, the inLowPerformance fluent is initiated by a lowPerformanceDetect-ed event and terminated by one of the events such as performanceNormalized or performanceNormFailed. Here the inLowPerformance event is activated when special AS-level performance service-level objectives (SLO) degrade (cf. Figure 6). Note that in ASSL, SLO are evaluated as *Booleans* based on their *satisfaction* and thus, they can be evaluated as *degraded* or *normal* [25]. Therefore, in our specification model, the lowPerformanceDetected event is activated anytime when the ADMARF's performance goes down. Alternatively, the performanceNormalized event activates when the same performance goes up.

As specified, the AS-level performance SLO are a global task whose realization is distributed among the AEs (DMARF stages). Thus, the AS-level performance degrades when the performance of any of the DMARF stages goes down (cf. the FOREACH loop in Figure 6), thus triggering the SELF_HEALING policy. In addition, the performanceNormFailed event activates if an a special event (selfHealingFailed) occurs in the system. This event is specified at the AE tier (cf. Section 3.1) and reports that the local AE-level self-healing has failed.

```
AESLO {
  SLO performance {
    METRICS.numberOfFailedNodes
    AND
    METRICS.numberOfProblematicNodes
  }
}
```

Fig. 7. AE Tier SLO

Although not presented in this specification, the `performanceNormFailed` event should be activated by any of the `performanceNormFailed` events specified for each AE (a DMARF stage).

Moreover, once the `inLowPerformance` fluent gets initiated, the corresponding `startSelfHealing` action is executed (cf. Figure 6). This action simply triggers an AE-level event if the performance of that AE is degraded. The AE-level event prompts the `SELF_HEALING` policy at the AE level (cf. Section 3.1).

AE Tier Specification for Self-Healing. At this tier we specify the self-healing policy for each AE in the ADMARF AS. Recall that the ADMARF 's AEs are the DMARF stages enriched with a special autonomic manager each.

Appendix A presents the self-healing specification of one AE called `STAGE_AE`. Note that the latter can be considered as a generic AE and the specifications of the four AEs (one per DMARF stage) can be derived from this one. Similar to the AS-level specification (cf. Section 3.1), here we specify a (but AE-level) `SELF_HEALING` policy with a set of *fluents* initiated and terminated by *events* and *actions* mapped to those *fluents* (cf. Appendix A). Thus we specified three distinct fluents: `inActiveSelfHealing`, `inFailedNodesDetected`, and `inProblematicNodesDetected`, each mapped to an AE-level action. The first fluent gets initiated when a `mustDoSelfHealing` event occurs in the system. That event is triggered by the AS-level `startSelfHealing` action in the case when the `performance` SLO of the AE get degraded (cf. Appendix A).

Here the `performance` SLO of the AE are specified as a Boolean expression over two ASSL metrics, such as the `numberOfFailedNodes` metric and the equivalent `numberOfProblematicNodes` (cf. Figure 7). Whereas the former measures the number of *failed notes* in the DMARF stage, the latter measures the number of *problematic nodes* in that stage.

Both metrics are specified as `RESOURCE` metrics, i.e., observing a managed resource controlled by the AE [25]. Note that the managed resource is the DMARF stage itself. Thus, as those metrics are specified (cf. Appendix A) they get updated by the DMARF stage via special *interface functions* embedded in the specification of a `STAGE_ME` managed element (cf. Appendix A). In addition, both metrics are set to accept only a zero value (cf. Figure 8), thus set in the so-called metric `THRESHOLD_CLASS` [25]. The latter determines rules for *valid* and *invalid* metric values. Since in ASSL metrics are evaluated as Booleans (valid or invalid) based on the value they are currently holding, the performance SLO (cf. Figure 7)

```
VALUE { 0 }
THRESHOLD_CLASS {     Integer [0] }  // valid only when holds 0
```

Fig. 8. AE Tier Metric Threshold Class

```
// runs the replica of a failed node
INTERFACE_FUNCTION runNodeReplica {
  PARAMETERS { DMARFNode  node }
  ONERR_TRIGGERS { EVENTS.nodeReplicaFailed }
}
 // recovers a problematic node
INTERFACE_FUNCTION recoverNode {
  PARAMETERS { DMARFNode  node }
  ONERR_TRIGGERS { EVENTS.nodeCannotBeFixed }
}
```

Fig. 9. AE Tier STAGE_ME Functions

gets degraded if one of the two defined metrics, the numberOfFailedNodes metric or the numberOfProblematicNodes metric become *invalid*, i.e., if the DMARF stage reports that there is one or more *failed* or *problematic* nodes.

The inActiveSelfHealing fluent prompts the analyzeProblem action execution (cf. Appendix A). The latter uses the STAGE_ME managed element's interface functions to determine the nature of the problem – is it a node that failed or it is a node that does not perform well. Based on this, the action triggers a mustSwitchToNodeReplica event or a mustFixNode event respectively. Each one of those events initiates a fluent in the AE SELF_HEALING policy to handle the performance problem. The inFailedNodesDetected fluent handles the case when a node has failed and its replica must be started and the inProblematicNodesDetected fluent handles the case when a node must be recovered. Here the first fluent prompts the execution of the startReplicaNode action and the second prompts the execution of the fixProblematicNode action. Internally, both actions call interface functions of the STAGE_ME managed element. Note that those functions trigger erroneous events if they do not succeed (cf. Figure 9). Those events terminate fluents of the AE SELF_HEALING policy (cf. Appendix A).

It is important to mention that the inFailedNodesDetected fluent gets initiated when the mustSwitchToNodeReplica event occurs in the system. The latter is triggered by the analyzeProblem action.

Moreover, the same event activates, according to its specification (cf. Figure 10), if a nodeCannotBeFixed event occurs in the system, which is due to the inability of the recoverNode interface function to recover the problematic node (cf. Figure 9). Therefore, if a node cannot be recovered the inFailedNodesDetec-ted fluent will be initiated in an attempt to start the replica of that node. Note that this conforms to the self-healing algorithm presented in Section 3.

```
EVENT mustSwitchToNodeReplica {
  ACTIVATION {
    OCCURRED {     EVENTS.nodeCannotBeFixed }
  }
}
```

Fig. 10. Event mustSwitchToNodeReplica

3.2   ASSL Self-Protection Model for DMARF

For scientific and research computing on a local network in a controlled lab environment, runs of DMARF do not need to be protected against malicious alteration or denial of service. However, as soon as the researchers across universities need to cooperate (or police departments to share the audio data recognition or computing), various needs about security and protection arise about data and computation results integrity, confidentiality, and the fact they come from a legitimate source. Therefore, *self-protection* of a DMARF-based system is less important in the localized scientific environments, but is a lot more important in global environments ran over the Internet, potentially through links crossing country borders. This is even more true if the data being worked on by a DMARF installation are of a sensitive nature such as recordings of the phone conversations of potential terrorist suspects. Thus, we point out the general requirements for this autonomic property of DMARF:

- For the self-protection aspect, the DMARF-based systems should adhere to the specification where each node proves its identity to other nodes participating in the pipeline as well as passive replicas. This will insure the data origin authentication (the data are coming from a legitimate source) and will protect against spoofing of the data with distorted voice recordings or incorrect processed data at the later stages of the pipeline. Thus, we ensure the trustworthiness of the distributed data being processed [6]. This can be achieved by proxy certificates issued to the nodes during the deployment and management phase. Each node digitally signs the outgoing data, with the signature the recipient node can verify through the certificate of the sender signed by trusted authority. This is in a way similar to how DNSSec [1] operates for the DNS names and servers by attaching a public key to the host and IP pair signed by the higher-level domain or authority. The similar trust mechanism is also important when DMARF is used for scientific research installation that say crosses the boundaries of several Universities' network perimeters over the Internet while performing scientific computation – the bottom line the data coming from the pipeline stages should be trustworthy, i.e. correct.
- The same proxy certificates can also help with the data privacy along public channels, especially when identities of real people are involved, such as speakers, that cross the Internet. The system should protect itself from any falsification attempt by detecting it, halting the corresponding computation, and logging and reporting the incident in a trustworthy manner.

- Run-time communication protocol selection is a self-protection option that ensures availability in case the default communication mechanism becomes unavailable (e.g. the default port of `rmiregistry` becomes blocked by a firewall) the participating nodes switch to XML-RPC over HTTP. The protection of self against Distributed Denial of Service (DDoS) is a difficult problem, which extends onto protecting not only self to the best available means but also the self's peers by avoiding flooding them by the self's own output of a compromised node. While the DDoS attacks are very difficult to mitigate if a node is under attack, each node can protect self and others by limiting the amount of outgoing traffic it, itself, produces when a compromise is suspected or too much traffic flood is detected.

For self-protection, DMARF-based systems should adhere to the specification where each node proves its identity to other nodes participating in the pipeline as well as passive replicas. This will insure the data origin authentication (the data are coming from a legitimate source) and will protect against spoofing of the data with distorted voice recordings or incorrect processed data at the later stages of the pipeline. Thus, we ensure the trustworthiness of the distributed data being processed [6]. This can be achieved by proxy certificates issued to the nodes during the deployment and management phase. Each node digitally signs the outgoing data, with the signature the recipient node can verify through the certificate of the sender signed by trusted authority. This is in a way similar to how DNSSec [1] operates for the DNS names and servers by attaching a public key to the host and IP pair signed by the higher-level domain or authority. The same proxy certificates can also help with the data privacy along public channels, especially when identities of real people are involved, such as speakers, that cross the Internet. The system should protect itself from any falsification attempt by detecting it, halting the corresponding computation, and logging and reporting the incident in a trustworthy manner.

To provide self-protecting capabilities, DMARF has to incorporate special autonomic computing behavior. To achieve that, similar to our related work on the self-healing and self-optimization models for DMARF [23,24], we add a special autonomic manager (AM) to each DMARF stage. This converts the latter into AEs that compose the autonomic DMARF (ADMARF) capable of self-management. Self-protecting is one of the self-management properties that must be addressed by ADMARF . Here we use ASSL to specify the self-protecting behavior of ADMARF where incoming messages must be secure in order to be able to process them. Thus, if a message (public or private) is about to be received in the AS, the following self-protection algorithm is followed by the AM of the stage (AE level) for private messages or by the global AM (AS level) for public messages:

- A message hook mechanism detects when a message (public or private) is about to be received.
- AM strives to identify the sender of that message by checking the embedded digital signature:

- If the message does not carry a digital signature then it is considered insecure.
- If the message carries a digital signature then its digital signature is checked:
  * If the digital signature is recognized then the message is considered secure.
  * If the digital signature is not recognized then the message is considered insecure.
  − If the message is secure no restrictions are imposed over the IO operations and the message can be processed further.
  − If the message is insecure the message is simply discarded by blocking any IO operations over that message.

The following sections describe the DMARF specification of the self-protecting algorithm revealed here. We specified this algorithm as an ASSL self-protecting policy spread on both system (AS tier) and autonomic element (AE tier) levels where *events*, *actions*, *metrics*, and special *managed element interface functions* are used to incorporate the self-protecting behavior in ADMARF (cf. Appendix B). In addition, two interaction protocols – a public (ASIP tier) and a private (AEIP tier), are specified to provide a secure communication system used by both DMARF nodes and external entities to communicate. Note that due to space limitations Appendix B presents a partial ASSL specification where only one AE (DMARF stage) is specified. The full specification specifies all the four DMARF stages.

IP Tiers Specification. Recall that ASSL specifies AEs as entities communicating via special *interaction protocols* (cf. Section 2.1). Note that all the communication activities (sending and receiving messages), all the communication channels, and all the communication entities (ASSL messages) must be specified in order to allow both internal and external entities to communicate. Hence, no entity can either send or receive a message that is not an ASSL-specified message or use alternative mechanism of communication. Thus, for the needs of the self-protecting mechanism, we specified two communication protocols – at the ASIP tier and at the AEIP tier (this is nested in the AE specification structure) (cf. Section 2.1). Please refer to Appendix B for a complete specification of both protocols.

At the ASIP tier, we specified a single public message (called `publicMessage`), a single sequential bidirectional public communication channel (called `publicLink`), and two public communication functions; specifically `receivePublicMessages` and `sendPublicMessages`. They are specified to receive and send public messages over the public channel. Here any message sent or received must be an instance of the ASSL-specified `publicMessage`. The latter has an embedded `ProxyCertificate` parameter specified to carry the digital signature of the message sender (cf. Appendix B). This parameter plays a key role in the self-protecting behavior of ADMARF . Every message sender must complete this parameter with its *proxy certificate* before sending the message or the latter will be discarded by the system.

```
//receive public messages if the message is secure
FUNCTION receivePublicMessages {
  DOES {
    IF ( AS.METRICS.thereIsInsecurePublicMessage ) THEN
      MESSAGES.publicMessage << CHANNELS.publicLink
    END
  }
}
```

Fig. 11. ASSL Specification of receivePublicMessages

```
SELF_PROTECTING {
  // a new incoming message has been detected
  FLUENT inSecurityCheck {
    INITIATED_BY { EVENTS.publicMessageIsComing }
    TERMINATED_BY {   EVENTS.publicMessageSecure,
                    EVENTS.publicMessageInsecure }
  }
  MAPPING {
    CONDITIONS { inSecurityCheck }
    DO_ACTIONS { ACTIONS.checkPublicMessage }
  }
}
```

Fig. 12. AS Tier SELF_PROTECTING Policy

Moreover, the mentioned communication functions (`receivePublicMessages` and `sendPublicMessages`) are the only ones specified in the entire AS able to process instances of the `publicMessage` ASSL message. Thus, to process such a message both functions are equipped with a conditional clause to check if the message is secure (cf. Figure 11).

Figure 11 shows the `receivePublicMessages` communication function. As is depicted, in order to send a public message the `thereIsInsecurePublicMessage` metric (cf. Appendix B) must be *valid*. The latter is updated by the self-protecting policy and is considered *invalid* (its operational evaluation returns `FALSE` [25]) if the public message to be received is *insecure*.

Note that the specification of the AEIP tier is identical to that of the ASIP tier (cf. Appendix B), but deals with private messages [25], i.e., external entities cannot send or receive such messages.

AS Tier Specification for Self-Protection. To protect the AS from insecure public messages we specified a self-management policy that handles the verification of any incoming public message. Thus, at this tier we specify a `SELF_PROTECTING` policy (one of the four self-CHOP policies [13]) to ensure protection from *insecure* public messages. Here we specified a single fluent mapped to an action via a mapping clause (cf. Figure 12).

The `inSecurityCheck` fluent is initiated by a `publicMessageIsComing` event and is terminated by one of the events, such as `publicMessageSecure` or `public-MessageInsecure`. Here the `inSecurityCheck` fluent is activated when an instance of the ASIP-specified `publicMessage` is sent to a recipient in the AS.

```
EVENTS { //these events are used in the fluents specification
  EVENT publicMessageIsComing {
    ACTIVATION { SENT { ASIP.MESSAGES.publicMessage } }
  }
  EVENT publicMessageInsecure {
    GUARDS { NOT METRICS.thereIsInsecurePublicMessage }
    ACTIVATION {
      CHANGED { METRICS.thereIsInsecurePublicMessage } }
  }
  EVENT publicMessageSecure {
    GUARDS { METRICS.thereIsInsecurePublicMessage }
    ACTIVATION {
      CHANGED { METRICS.thereIsInsecurePublicMessage } }
  }
}  // EVENTS
```

Fig. 13. AS Tier Events

Recall that any public message to be sent to a system recipient (e.g., a DMARF node) must be an instance of the ASSL `publicMessage` message (cf. Section 3.2). Therefore, in our specification model, the `publicMessageSecure` event will be activated anytime when a `publicMessage` is about to be received by the AS (by a recipient in that AS).

Figure 13 presents the specification of all the three events used to initiate and terminate the `inSecurityCheck` fluent. As it is depicted, both `publicMessageInsecure` and `publicMessageSecure` are prompted when `thereIsInsecurePublicMessage`'s value has changed. Special GUARDS are specified to prevent those events be prompted when that metric is *valid* or *not valid* respectively [25]. The corresponding metric `thereIsInsecurePublicMessage` accepts only Boolean values and is valid when it holds FALSE. The same metric is set to TRUE or FALSE by the `checkPublicMessage` action. Here the metric is set to TRUE anytime when a new public insecure message has been discovered (cf. Appendix B).

The `checkPublicMessage` action is mapped to the `inSecurityCheck` fluent (cf. Figure 12). Here this action is performed anytime when the AS enters in a *security check* state (determined by the `inSecurityCheck` fluent). This action is intended to check how secure is the incoming `publicMessage`, which triggers the self-protecting policy by prompting the `publicMessageIsComing` event (cf. Figure 13). To do that, the `checkPublicMessage` action calls for each AE in the AS a `checkSenderCertificate` action that must be specified in each AE (DMARF stage) (cf. Figure 14).

The `checkSenderCertificate` action returns TRUE if the `publicMessage` carries a valid digital signature (cf. Appendix B), i.e., the message is sent by a trusted sender (DMARF node). As depicted by Figure 14, if one of the AE returns TRUE, then the `publicMessage` is considered secure; otherwise, it is considered insecure. If the message is insecure the `thereIsInsecurePublicMessage` metric is set to TRUE, which blocks the IO operations over this message (cf. Section 3.2 and Appendix B).

```
senderIdentified = false;
FOREACH member in AES {
  IF ( NOT senderIdentified ) THEN
    senderIdentified =
          call member.ACTIONS.checkSenderCertificate
          (ASIP.MESSAGES.publicMessage.senderSignature)
  END
};
IF NOT senderIdentified THEN
  // makes the metric invalid and thus, triggers the attached
  // event and blocks all the operations with public messages
  set METRICS.thereIsInsecurePublicMessage.VALUE = true
END
```

Fig. 14. AS checkPublicMessage Action – Partial Specification

```
MANAGED_ELEMENTS {
  MANAGED_ELEMENT STAGE_ME {
    // checks if a node certificate is valid
    INTERFACE_FUNCTION checkNodeCertificate {
      PARAMETERS { ProxyCertificate    theCertificate }
      RETURNS { Boolean }
    }
  }
}
```

Fig. 15. AE STAGE_ME Managed Element

AE Tier Specification for Self-Protection. At this tier we specify the self-protecting mechanism for private messages. Thus, for each AE (DMARF stage) we specify a SELF_PROTECTING self-management policy identical to the same policy specified at the AS tier (cf. Section 3.2). Note that this policy deals with private messages specified at the AEIP tier (the AE's private interaction protocol – cf. Appendix B).

Therefore, similarly to the same policy specified for the AS tier, the AE-level SELF_PROTECTING policy is specified with a single inSecurityCheck fluent mapped to a checkPrivateMessage action. The inSecurityCheck fluent is initiated by the privateMessageIsComming event and terminated by the privateMessageIsSecure event or by the privateMessageSecure event. These events are similar to their homologous events specified at the AS tier (cf. Section 3.2), but dealing with the AEIP-specified privateMessage message and with the thereIsInsecurePrivateMessage metric at the AE-level (cf. Appendix B).

To perform the security checks of incoming private messages, the checkPrivateMessage action invokes the checkSenderCertificate action (recall that the same action is called by the checkPublicMessage action to check public messages – cf. Section 3.2). Internally, the checkSenderCertificate action calls a *managed element interface function* specified at the AEIP protocol to check proxy certificates (cf. Figure 15).

Recall (cf. Section 2.1) that managed elements provide special interface functions to control the DMARF system. Hence, as depicted by Figure 15, we expect the DMARF stage to verify whether a specific proxy certificate is valid and to return TRUE or FALSE. DMARF does that through the Java Data Security Framework (JDSF) [9,11].

3.3   ASSL Self-Optimization Model for DMARF

The two major functional requirements applicable to large DMARF installations related to self-optimization are outlined below:

Training set classification data replication. A DMARF-based system may do a lot of multimedia data processing and number crunching throughout the pipeline. The bulk of I/O-bound data processing falls on the sample loading stage and the classification stage. The preprocessing, feature extraction, and classification stages also do a lot of CPU-bound number crunching, matrix operations, and other potentially heavy computations. The stand-alone local MARF instance employs dynamic programming to cache intermediate results, usually as feature vectors, inverse co-variance matrices, and other array-like data. A lot of this data is absorbed by the classification stages. In the case of the DMARF, such data may end up being stored on different hosts that run the classification service potentially causing re-computation of the already computed data on other classification host that did a similar evaluation already. Thus, the classification stage nodes need to communicate to exchange the data they have lazily acquired among all the classification members. Such data mirroring/replication would optimize a lot of computational effort on the end nodes.

Dynamic communication protocol selection. Additional aspect of self-optimization is automatic selection of the available most efficient communication protocol. E.g., if DMARF initially uses WebServices XML-RPC and later discovers all of its nodes can also communicate using say Java RMI, they can switch to that as their default protocol in order to avoid marshaling and de-marshaling heavy SOAP XML messages that are always a subject of a big overhead even in the compressed form.

Here, the DMARF Classification stage is augmented with a self-optimizing autonomic policy. We used ASSL to specify this policy and generate implementation for the same. Appendix C presents a partial specification of the ASSL self-optimization model for ADMARF. As specified, the autonomic behavior is encoded in a special ASSL construct denoted as SELF_OPTIMIZING policy. The latter is specified at two levels - the global AS-tier level and the level of single AE (the AE-tier). The algorithm behind is described by the following elements:

- Any time when ADMARF enters in the Classification stage, a self-optimization behavior takes place.
- The Classification stage itself forces the stage nodes synchronize their latest cached results. Here each node is asked to get the results of the other nodes.
- Before proceeding with the problem computation, each stage node strives to adapt to the most efficient currently available communication protocol.

```
SELF_OPTIMIZING {
  // DMARF enters in the Classification Stage
  FLUENT inClassificationStage {
    INITIATED_BY { EVENTS.enteringClassificationStage }
    TERMINATED_BY {    EVENTS.optimizationSucceeded,
                       EVENTS.optimizationNotSucceeded }
  }
  MAPPING {
    CONDITIONS { inClassificationStage }
    DO_ACTIONS { ACTIONS.runGlobalOptimization }
  }
}
```

Fig. 16. AS Tier SELF-OPTIMIZING Policy

The following sections describe the ASSL specification of the self-optimization algorithm revealed here.

AS Tier Specification for Self-Optimization. At this tier we specify a system-level `SELF_OPTIMIZING` policy and the actions and events supporting that policy. As was mentioned, ASSL supports policy specification with special constructs called *fluents* and *mappings* [25]. Whereas the former are special states with conditional duration, the latter map actions to be executed when the system enters in such a state.

Figure 16 depicts the AS-tier specification of the `SELF_OPTIMIZING` policy. As we see the policy is triggered when the special fluent `inClassificationStage` is initiated. Here when ADMARF enters the Classification stage in its pipeline, an AS-level `enteringClassificationStage` event is prompted to initiate the corresponding `inClassificationStage` fluent.

Further, this fluent is mapped to an AS-level `runGlobalOptimization` action (cf. Appendix C). This action iterates over all the Classification stage nodes specified as distinct AEs (cf. Section 3.3) and calls for each node a special AE-level `synchronizeResults` action (cf. Appendix C). In case of exception, the `optimizationNotSucceeded` event is issued; else the `optimizationSucceeded` event is issued. Both events terminate the `inClassificationStage` fluent, and consecutively ADMARF exits the `SELF_OPTIMIZING` policy.

To distinguish the AEs from the other AEs in ADMARF, we specified the architecture topology of the system. For this we used the `ASARCHITECTURE` ASSL construct [25]. Appendix C presents the specification of the ADMARF architecture topology. Note that this is a partial specification depicting only two AEs. The full `ASARCHITECTURE` specification includes all the AEs of ADMARF. As depicted, we specified a special group of AEs called `CLASSF_STAGE` with members all the AEs representing the Classification stage nodes. This group allows the `runGlobalOptimization` action iterates over the stage nodes.

AE Tier Specification for Self-Optimization. At this tier we specified the `SELF_OPTIMIZING` policy for the Classification stage nodes. Here we specified for every node a distinct AE. (cf. Appendix C) presents the partial specification of two AEs, each representing a single node of the Specification stage. At this level, self-optimization concentrates on adapting the single nodes to the most efficient communication protocol. Similar to the AS-level policy specification (cf. Section 3.3), an inCPAdaptation fluent is specified to trigger such adaptation when ADMARF enters in the Specification stage. This fluent is initiated by the AS-level enteringClassificationStage event.

The same fluent is mapped to an adaptCP action to perform the needed adaptation. This action is specified as IMPL, i.e., requiring further implementation [25]. In ASSL, we specify IMPL actions to hide complexity via abstraction. Here, the adaptCP action is a complex structure, which explanation is beyond the scope of this paper. Therefore, we abstracted the specification of this action (through IMPL) and provided only the prerequisite *guard* conditions and prompted events.

4 Conclusion

In this article, we have presented ASSL specification models for autonomic features of ADMARF. To develop these features, we devised algorithms with ASSL for the pipelined stages of the DMARF's pattern recognition pipeline. The autonomic features were specified as special self-managing policies for self-healing, self-protecting, and self-optimizing in ADMARF. The ADMARF system (upon completion of the open-source implementation) will be able to fully function in autonomous environments, be those on the Internet, large multimedia processing farms, robotic spacecraft that do their own analysis, or simply even pattern-recognition research groups that can rely more on the availability of their systems that run for multiple days, unattended. Although not a fully complete specification model for ADMARF, we have attempted to provide didactic evidence of how ASSL can help us achieve desired automicity in DMARF.

Future work is concerned with further ADMARF development by including new autonomic features. For example, together with the full implementation and testing of the presented specification models, we intend to develop autonomic features covering the self-configuration aspects of ADMARF. These will help to construct an intelligent ADMARF system able to react automatically to dynamic requirements by finding possible solutions and applying those with no human interaction.

References

1. DNSSEC.NET. DNSSEC: DNS Security Extensions Securing the Domain Name System (2002-2010), http://www.dnssec.net/ (last viewed November 2010)
2. Havelund, K., Pressburger, T.: Model checking Java programs using Java PathFinder. STTT 2(4), 366–381 (2000)
3. Horn, P.: Autonomic computing: IBM's perspective on the state of information technology. Technical report, IBM T. J. Watson Laboratory (October 2001)
4. IBM Corporation. An architectural blueprint for autonomic computing. Technical report, IBM Corporation (2006)
5. Mokhov, S.A.: Choosing Best Algorithm Combinations for Speech Processing Tasks in Machine Learning Using MARF. In: Bergler, S. (ed.) Canadian AI 2008. LNCS (LNAI), vol. 5032, pp. 216–221. Springer, Heidelberg (2008)

6. Mokhov, S.A.: Towards security hardening of scientific distributed demand-driven and pipelined computing systems. In: Proceedings of the 7th International Symposium on Parallel and Distributed Computing (ISPDC 2008), pp. 375–382. IEEE Computer Society (July 2008)

7. Mokhov, S.A.: Towards syntax and semantics of hierarchical contexts in multimedia processing applications using MARFL. In: Proceedings of the 32nd Annual IEEE International Computer Software and Applications Conference (COMPSAC), Turku, Finland, pp. 1288–1294. IEEE Computer Society (July 2008)

8. Mokhov, S.A., Debbabi, M.: File type analysis using signal processing techniques and machine learning vs. file unix utility for forensic analysis. In: Goebel, O., Frings, S., Guenther, D., Nedon, J., Schadt, D. (eds.) Proceedings of the IT Incident Management and IT Forensics (IMF 2008), Mannheim, Germany. LNI, vol. 140, pp. 73–85. GI (2008)

9. Mokhov, S.A., Huynh, L.W., Wang, L.: The integrity framework within the Java Data Security Framework (JDSF): Design refinement and implementation. In: Sobh, T., Elleithy, K., Mahmood, A. (eds.) Novel Algorithms and Techniques in Telecommunications and Networking, Proceedings of CISSE 2008, pp. 449–455. Springer, Heidelberg (December 2008), printed in (January 2010)

10. Mokhov, S.A., Jayakumar, R.: Distributed Modular Audio Recognition Framework (DMARF) and its applications over web services. In: Sobh, T., Elleithy, K., Mahmood, A. (eds.) Proceedings of TeNe 2008, University of Bridgeport, CT, USA, pp. 417–422. Springer, Heidelberg (December 2008), printed in (January 2010)

11. Mokhov, S.A., Rassai, F., Huynh, L.W., Wang, L.: The authentication framework within the Java Data Security Framework (JDSF): Design refinement and implementation. In: Sobh, T., Elleithy, K., Mahmood, A. (eds.) Novel Algorithms and Techniques in Telecommunications and Networking, Proceedings of CISSE 2008, pp. 423–429. Springer, Heidelberg (2008), printed in (January 2010)

12. Mokhov, S.A., Vassev, E.: Autonomic specification of self-protection for Distributed MARF with ASSL. In: Proceedings of C3S2E 2009, pp. 175–183. ACM, New York (2009)

13. Murch, R.: Autonomic Computing: On Demand Series. IBM Press, Prentice Hall (2004)

14. Sun Microsystems, Inc. Java IDL. Sun Microsystems, Inc. (2004)

15. Sun Microsystems, Inc. The Java web services tutorial (for Java Web Services Developer's Pack, v2.0). Sun Microsystems, Inc. (February 2006)

16. The Planetary Society. Space topics: Voyager - the story of the mission (2009), http://planetary.org/explore/topics/spacemissions/voyager/objectives.html

17. Truszkowski, W., Hinchey, M., Rash, J., Rouff, C.: NASA's swarm missions: The challenge of building autonomous software. IT Professional 6(5), 47–52 (2004)

18. Vassev, E.: ASSL: Autonomic System Specification Language – A Framework for Specification and Code Generation of Autonomic Systems. LAP Lambert Academic Publishing (November 2009) ISBN: 3-838-31383-6

19. Vassev, E., Hinchey, M.G., Paquet, J.: Towards an ASSL specification model for NASA swarm-based exploration missions. In: Proceedings of the 23rd Annual ACM Symposium on Applied Computing (SAC 2008) - AC Track, pp. 1652–1657. ACM (2008)

20. Vassev, E., Hinchey, M.: ASSL specification and code generation of self-healing behavior for NASA swarm-based systems. In: Proceedings of the 6th IEEE International Workshop on Engineering of Autonomic and Autonomous Systems (EASe 2009), pp. 77–86. IEEE Computer Society (2009)

21. Vassev, E., Hinchey, M.: Modeling the image-processing behavior of the NASA Voyager mission with ASSL. In: Proceedings of the 3rd IEEE International Conference on Space Mission Challenges for Information Technology (SMC-IT 2009), pp. 246–253. IEEE Computer Society (2009)

22. Vassev, E., Hinchey, M., Paquet, J.: A self-scheduling model for NASA swarm-based exploration missions using ASSL. In: Proceedings of the Fifth IEEE International Workshop on Engineering of Autonomic and Autonomous Systems (EASe 2008), pp. 54–64. IEEE Computer Society (2008)

23. Vassev, E., Mokhov, S.A.: Self-optimization property in autonomic specification of Distributed MARF with ASSL. In: Shishkov, B., Cordeiro, J., Ranchordas, A. (eds.) Proceedings of ICSOFT 2009, Sofia, Bulgaria, vol. 1, pp. 331–335. INSTICC Press (July 2009)

24. Vassev, E., Mokhov, S.A.: Towards Autonomic Specification of Distributed MARF with ASSL: Self-Healing. In: Lee, R., Ormandjieva, O., Abran, A., Constantinides, C. (eds.) SERA 2010. SCI, vol. 296, pp. 1–15. Springer, Heidelberg (2010)

25. Vassev, E.I.: Towards a Framework for Specification and Code Generation of Autonomic Systems. PhD thesis, Department of Computer Science and Software Engineering, Concordia University, Montreal, Canada (2008)

26. Wollrath, A., Waldo, J.: Java RMI tutorial. Sun Microsystems, Inc. (1995-2005)

A   ASSL Specification of DMARF Self-Healing

```
// ASSL self-healing specification model for DMARF
AS DMARF {

  TYPES { DMARFNode }

  ASSLO {
    SLO performance {....}
  }

  ASSELF_MANAGEMENT {
    SELF_HEALING {....}
  } // ASSELF_MANAGEMENT

  ACTIONS {
    ACTION IMPL startSelfHealing {
      GUARDS { ASSELF_MANAGEMENT.SELF_HEALING.inLowPerformance }
      TRIGGERS {
        IF NOT AES.STAGE_AE.AESLO.performance THEN
          AES.STAGE_AE.EVENTS.mustDoSelfHealing
        END
      }
    }
  } // ACTIONS

  EVENTS { // these events are used in the fluents specification
    EVENT lowPerformanceDetected { .... }
    EVENT performanceNormalized { .... }
    EVENT performanceNormFailed { .... }
  } // EVENTS

} // AS DMARF

AES {

  AE STAGE_AE {

    VARS { DMARFNode nodeToRecover }

    AESLO {
      SLO performance {....}
    }

    AESELF_MANAGEMENT {
      SELF_HEALING {
        FLUENT inActiveSelfHealing {
          INITIATED_BY { EVENTS.mustDoSelfHealing }
          TERMINATED_BY { EVENTS.selfHealingSuccessful,
            EVENTS.selfHealingFailed }
        }
        FLUENT inFailedNodesDetected {
```

```
                  INITIATED_BY { EVENTS.mustSwitchToNodeReplica }
                  TERMINATED_BY { EVENTS.nodeReplicaStarted,
                    EVENTS.nodeReplicaFailed }
                }
                FLUENT inProblematicNodesDetected {
                  INITIATED_BY { EVENTS.mustFixNode }
                  TERMINATED_BY { EVENTS.nodeFixed,
                    EVENTS.nodeCannotBeFixed }
                }
                MAPPING {
                  CONDITIONS { inActiveSelfHealing }
                  DO_ACTIONS { ACTIONS.analyzeProblem }
                }
                MAPPING {
                  CONDITIONS { inFailedNodesDetected }
                  DO_ACTIONS { ACTIONS.startReplicaNode }
                }
                MAPPING {
                  CONDITIONS { inProblematicNodesDetected }
                  DO_ACTIONS { ACTIONS.fixProblematicNode }
                }
              }
          } // AESELF_MANAGEMENT

          AEIP {
            MANAGED_ELEMENTS {
              MANAGED_ELEMENT STAGE_ME {
                INTERFACE_FUNCTION countFailedNodes {....}
                INTERFACE_FUNCTION countProblematicNodes {....}
                // returns the next failed node
                INTERFACE_FUNCTION getFailedNode {....}
                // returns the next problematic node
                INTERFACE_FUNCTION getProblematicNode {....}
                // runs the replica of a failed nodee
                INTERFACE_FUNCTION runNodeReplica {....}
                // recovers a problematic node
                INTERFACE_FUNCTION recoverNode {....}
              }
            }
          } // AEIP

          ACTIONS {
            ACTION analyzeProblem {
              GUARDS { AESELF_MANAGEMENT.SELF_HEALING.inActiveSelfHealing }
              VARS { BOOLEAN failed }
              DOES {
                IF METRICS.numberOfFailedNodes THEN
                  AES.STAGE_AE.nodeToRecover =
            call AEIP.MANAGED_ELEMENTS.STAGE_ME.getFailedNode;
                  failed = TRUE
                END
                ELSE
                  AES.STAGE_AE.nodeToRecover =
            call AEIP.MANAGED_ELEMENTS.STAGE_ME.getProblematicNode;
                  failed = FALSE
                END
              }
              TRIGGERS {
                IF failed THEN EVENTS.mustSwitchToNodeReplica END
                ELSE EVENTS.mustFixNode END
              }
              ONERR_TRIGGERS { EVENTS.selfHealingFailed }
            }

            ACTION startReplicaNode {
              GUARDS { AESELF_MANAGEMENT.SELF_HEALING.inFailedNodesDetected }
              DOES {
call AEIP.MANAGED_ELEMENTS.STAGE_ME.runNodeReplica(AES.STAGE_AE.nodeToRecover) }
              TRIGGERS {   EVENTS.nodeReplicaStarted }
            }

            ACTION fixProblematicNode {
              GUARDS { AESELF_MANAGEMENT.SELF_HEALING.inProblematicNodesDetected }
              DOES {
call AEIP.MANAGED_ELEMENTS.STAGE_ME.recoverNode(AES.STAGE_AE.nodeToRecover) }
              TRIGGERS {   EVENTS.nodeFixed}
            }
          } // ACTIONS

          EVENTS {
            EVENT mustDoSelfHealing { }
            EVENT selfHealingSuccessful {
              ACTIVATION {
                OCCURRED {   EVENTS.nodeReplicaStarted }
                OR
                OCCURRED {   EVENTS.nodeFixed }
              }
            }
            EVENT selfHealingFailed {
              ACTIVATION {
```

```
            OCCURRED {   EVENTS.nodeReplicaFailed }
        }
    }
    EVENT mustSwitchToNodeReplica {....}
    EVENT nodeReplicaStarted { }
    EVENT nodeReplicaFailed { }
    EVENT mustFixNode { }
    EVENT nodeFixed {  }
    EVENT nodeCannotBeFixed { }
    } // EVENTS

    METRICS {
      // increments when a failed node has been discovered
      METRIC numberOfFailedNodes {
        METRIC_TYPE { RESOURCE  }
        METRIC_SOURCE {   AEIP.MANAGED_ELEMENTS.STAGE_ME.countFailedNodes }
        DESCRIPTION {"counts failed nodes in the MARF stage"}
        VALUE { 0 }
        THRESHOLD_CLASS {  Integer [0] } // valid only when holding 0 value
      }
      // increments when a problematic node has been discovered
      METRIC numberOfProblematicNodes {
        METRIC_TYPE { RESOURCE  }
        METRIC_SOURCE {   AEIP.MANAGED_ELEMENTS.STAGE_ME.countProblematicNodes }
        DESCRIPTION {"counts nodes with problems in the MARF stage"}
        VALUE { 0 }
        THRESHOLD_CLASS {  Integer [0] } // valid only when holding 0 value
      }
    }
  }
}
```

B   ASSL Code Specification for DMARF Self-Protection

```
// ASSL self-protecting specification model for DMARF
AS DMARF {

  TYPES { ProxyCertificate }

  ASSELF_MANAGEMENT {
    // if a private message is detected as being insecure then
    // ignore it - the AE cannot neither receive nor resend it
    SELF_PROTECTING {....}

  } // ASSELF_MANAGEMENT

  ACTIONS {
    ACTION checkPublicMessage {
      GUARDS { ASSELF_MANAGEMENT.SELF_PROTECTING.inSecurityCheck }
      VARS { Boolean senderIdentified }
      DOES {
        senderIdentified = false;
        FOREACH member in AES {
          IF ( NOT senderIdentified ) THEN
            senderIdentified =
              call member.ACTIONS.checkSenderCertificate
              (ASIP.MESSAGES.publicMessage.senderSignature)
          END
        };
        IF NOT senderIdentified THEN
// makes the metric invalid and thus, triggers the attached event
// and blocks all the operations with public messages
          set METRICS.thereIsInsecurePublicMessage.VALUE = true
        END
        ELSE
// makes the metric valid and thus, triggers the attached event
// and unblocks all the operations with public messages
          set METRICS.thereIsInsecurePublicMessage.VALUE = false
        END
      }
      ONERR_DOES {
        // if error then treat the message as insecure
        set METRICS.thereIsInsecurePublicMessage.VALUE = true
      }
    }
  } // ACTIONS

  EVENTS { // these events are used in the fluents specification
    EVENT publicMessageIsComing {....}
    EVENT publicMessageInsecure {....}
    EVENT publicMessageSecure {....}
  } // EVENTS

  METRICS {
    // set to true when a new public insecure message
    // has been discovered
    METRIC thereIsInsecurePublicMessage {
      METRIC_TYPE { QUALITY  }
```

```
          DESCRIPTION {"detects an insecure message in the AE"}
          VALUE { false }
          // valid only when holding false value
          THRESHOLD_CLASS { Boolean [false] }
        }
      }
  } // AS DMARF

  ASIP {
    MESSAGES {
      MESSAGE publicMessage {....}
    }

    CHANNELS {
      CHANNEL publicLink {....}
    }

    FUNCTIONS {
      //receive public messages if the message is secure
      FUNCTION receivePublicMessages {....}
      //send public messages if the message is secure
      FUNCTION sendPublicMessages {....}
    }
  }

  AES {

    AE STAGE_AE {

      AESELF_MANAGEMENT {
        // if a private message is detected as being insecure then
        // ignore it - the AE cannot neither receive nor resend it
        SELF_PROTECTING {
          FLUENT inSecurityCheck {
            INITIATED_BY { EVENTS.privateMessageIsComming }
            TERMINATED_BY { EVENTS.privateMessageSecure,
                            EVENTS.privateMessageInsecure }
          }
          MAPPING {
            CONDITIONS { inSecurityCheck }
            DO_ACTIONS { ACTIONS.checkPrivateMessage }
          }
        }
      } // AESELF_MANAGEMENT

      AEIP {
        MESSAGES {
          MESSAGE privateMessage {....}
        }

        CHANNELS {
          CHANNEL privateLink {....}
        }

        FUNCTIONS {
          //receive private messages if the message is secure
          FUNCTION receivePrivateMessages {
            DOES {
              IF ( AES.STAGE_AE.METRICS.thereIsInsecurePrivateMessage ) THEN
                AEIP.MESSAGES.privateMessage << AEIP.CHANNELS.privateLink
              END
            }
          }
          //send private messages if the message is secure
          FUNCTION sendPrivateMessages {
            DOES {
              IF ( AES.STAGE_AE.METRICS.thereIsInsecurePrivateMessage ) THEN
                AEIP.MESSAGES.privateMessage >> AEIP.CHANNELS.privateLink
              END
            }
          }
        }

        MANAGED_ELEMENTS {
          MANAGED_ELEMENT STAGE_ME {....}
        }
      } // AEIP

      ACTIONS {
        ACTION checkSenderCertificate {
          PARAMETERS { ProxyCertificate theCertificate }
          RETURNS { Boolean }
          VARS { Boolean found }
          DOES {
            found = call AEIP.MANAGED_ELEMENTS.STAGE_ME.checkNodeCertificate
              (theCertificate);
            return found
          }
        }
```

```
ACTION checkPrivateMessage {
  GUARDS { AESELF_MANAGEMENT.SELF_PROTECTING.inSecurityCheck }
  VARS { Boolean senderIdentified }
  DOES {
    senderIdentified = call ACTIONS.checkSenderCertificate
      ( AEIP.MESSAGES.privateMessage.senderSignature );
    IF NOT senderIdentified THEN
      // makes the metric invalid and thus, triggers the attached event
      // and blocks all the operations with private messages
      set METRICS.thereIsInsecurePrivateMessage.VALUE = true
    END
    ELSE
      // makes the metric valid and thus, triggers the attached event
      // and unblocks all the operations with private messages
      set METRICS.thereIsInsecurePrivateMessage.VALUE = false
    END
  }
  ONERR_DOES {
    // if error then treat the message as insecure
    set METRICS.thereIsInsecurePrivateMessage.VALUE = true
  }
}
} // ACTIONS

EVENTS {
  EVENT privateMessageIsComming {
    ACTIVATION { SENT { AEIP.MESSAGES.privateMessage } }
  }
  EVENT privateMessageInsecure {
    GUARDS { NOT METRICS.thereIsInsecurePrivateMessage }
    ACTIVATION { CHANGED { METRICS.thereIsInsecurePrivateMessage } }
  }
  EVENT privateMessageSecure {
    GUARDS { METRICS.thereIsInsecurePrivateMessage }
    ACTIVATION { CHANGED { METRICS.thereIsInsecurePrivateMessage } }
  }
} // EVENTS

METRICS {
  // set to true when an insecure private message is discovered
  METRIC thereIsInsecurePrivateMessage {
    METRIC_TYPE { QUALITY }
    DESCRIPTION {"detects an insecure message in the AE"}
    VALUE { false }
    // valid only when holding false value
    THRESHOLD_CLASS { Boolean [false] }
  }
}
}
}
```

C   ASSL Code Specification for DMARF Self-Optimization

```
// ASSL self-optimization specification model for DMARF
AS DMARF {

  ASELF_MANAGEMENT {
    // DMARF strives to optimize by synchronizing cached
    // results before starting with the Classification Stage
    SELF_OPTIMIZING {....}
  } // ASELF_MANAGEMENT

  ASARCHITECTURE {....}

  ACTIONS {
    ACTION runGlobalOptimization {
      GUARDS { ASELF_MANAGEMENT.SELF_OPTIMIZING.inClassificationStage }
      DOES {
        FOREACH member IN ASARCHITECTURE.GROUPS.CLASSF_STAGE.MEMBERS {
          call IMPL member.ACTIONS.synchronizeResults
        }
      }
      TRIGGERS {
        EVENTS.optimizationSucceeded
      }
      ONERR_TRIGGERS {
        // if error then report unsuccessful optimization
        EVENTS.optimizationNotSucceeded
      }
    }
  } // ACTIONS

  EVENTS { // these events are used in the fluents specification
    EVENT enteringClassificationStage { }
    EVENT optimizationSucceeded { }
    EVENT optimizationNotSucceeded { }
```

```
    } // EVENTS

} // AS DMARF

AES {

  AE CLASSF_STAGE_NODE_1 {

    AESELF_MANAGEMENT {
      SELF_OPTIMIZING {
        FLUENT inCPAdaptation {
          INITIATED_BY { AS.EVENTS.enteringClassificationStage }
          TERMINATED_BY {  EVENTS.cpAdaptationSucceeded,
              EVENTS.cpAdaptationNotSucceeded }
        }
        MAPPING {
          CONDITIONS { inCPAdaptation }
          DO_ACTIONS { ACTIONS.adaptCP }
        }
      }
    }

    ACTIONS {
      ACTION IMPL synchronizeResults {
        GUARDS { AS.ASSELF_MANAGEMENT.SELF_OPTIMIZING.
            inClassificationStage
        }
      }
      ACTION IMPL adaptCP {
        GUARDS { AESELF_MANAGEMENT.SELF_OPTIMIZING.inCPAdaptation }
        TRIGGERS { EVENTS.cpAdaptationSucceeded }
        ONERR_TRIGGERS { EVENTS.cpAdaptationNotSucceeded }
      }
    } // ACTIONS

    EVENTS { // these events are used in the fluents specification
      EVENT cpAdaptationSucceeded { }
      EVENT cpAdaptationNotSucceeded { }
    } // EVENTS
  }

  AE CLASSF_STAGE_NODE_2 {

    AESELF_MANAGEMENT {
      SELF_OPTIMIZING {
        FLUENT inCPAdaptation {
          INITIATED_BY { AS.EVENTS.enteringClassificationStage }
          TERMINATED_BY {  EVENTS.cpAdaptationSucceeded,
              EVENTS.cpAdaptationNotSucceeded }
        }
        MAPPING {
          CONDITIONS { inCPAdaptation }
          DO_ACTIONS { ACTIONS.adaptCP }
        }
      }
    }

    ACTIONS {
      ACTION IMPL synchronizeResults {
        GUARDS { AS.ASSELF_MANAGEMENT.SELF_OPTIMIZING.
            inClassificationStage
        }
      }
      ACTION IMPL adaptCP {
        GUARDS { AESELF_MANAGEMENT.SELF_OPTIMIZING.inCPAdaptation }
        TRIGGERS { EVENTS.cpAdaptationSucceeded }
        ONERR_TRIGGERS { EVENTS.cpAdaptationNotSucceeded }
      }
    } // ACTIONS

    EVENTS { // these events are used in the fluents specification
      EVENT cpAdaptationSucceeded { }
      EVENT cpAdaptationNotSucceeded { }
    } // EVENTS}
  }
}
```

Autonomic Nature-Inspired Eco-systems

Antonio Manzalini[1], Nermin Brgulja[2],
Corrado Moiso[1], and Roberto Minerva[1,3]

[1] Telecom Italia, Strategy and Innovation, Torino, Italy
{antonio.manzalini,corrado.moiso,roberto.minerva}@telecomitalia.it
[2] University of Kassel, Kassel, Germany
nermin.brgulja@comtec.eecs.uni-kassel.de
[3] Institut TELECOM SudParis, Evry, France

Abstract. Computing and storage powers are progressively embedded in all sort of electronic devices interconnected by ubiquitous communications. This creates pervasive and complex environments fostering the development and evolution of services eco-systems. Autonomic technologies coupled with additional bio-inspired principles can provide strong elements for facing such challenge. This paper aims at looking inside the black box of an autonomic bio-inspired eco-system. Specifically, a model of autonomic component is elaborated allowing the evolution of eco-systems enabled by means of self-awareness and self-organization. This approach goes beyond a traditional mechanistic one, where concepts derived from biology are applied to explain what happens in an eco-system: in fact, behaviour and evolution of a service eco-system are engineered by programming the components. This model has been implemented and experimentally validated within the EU project CASCADAS, so some validation results as also reported. Moreover, the paper analyses a use case concerning a decentralized server farm, being considered part of a complex eco-system.

Keywords: Autonomic Computing and Communication, Self-Organization, Service Eco-systems, Bio-inspired primitives.

1 Introduction

Todays networks are becoming increasingly ubiquitous, complex and heterogeneous: computing and storage resources are embedded in different types of nodes and devices that are interconnected through a variety of wired and wireless networks and protocols. People, smart objects, machines and the surrounding space (e.g., sensors, RFID tags, etc.) will create a pervasive and complex computational and networking environment fostering the evolution towards eco-systems of resource, services and data. On the other hand, dynamicity and ubiquity of said future environments will determine on the way new challenges and requirements which cannot be satisfied by current solutions, as analysed in Section 3. Autonomic technologies coupled with additional bio-inspired primitives, such as gossip, random choice, fields, and gradients, can provide strong elements to foster this evolution, overcoming the main obstacles to the exploitation of pervasive

M.L. Gavrilova et al. (Eds.): Trans. on Comput. Sci. XV, LNCS 7050, pp. 158–191, 2012.
© Springer-Verlag Berlin Heidelberg 2012

and complex computational and networking environments. As a matter of fact, this is an avenue of considerable research and industrial interest [39].

The goal of this paper is to open the black box of an autonomic bio-inspired eco-system and to come to some understanding of how it generates its dynamic behaviours. This paper elaborates a model of autonomic component whose self-awareness enables the evolution of the eco-system through adaptation and self-organization emerges by using bio-inspired interactions and cooperation. This approach goes beyond a traditional mechanistic one, where concepts derived from biology are applied just to describe the evolution of a complex eco-system, and not used to engineer its behaviour: in fact, proposed approach has the goal of engineering a service eco-system by programming its components.

Specifically, the paper is presenting the vision and the main concepts of a model of an autonomic component autonomic model and of the associated Toolkit developed in the EU project CASCADAS [32], [36]. The main goal of the project has been defining, developing and experimentally validating a novel architectural paradigm of service eco-systems for future ICT-systems (i.e. Telecommunications and Internet networks and systems). The architecture of CASCADAS eco-systems is based on a pervasive environment of distributed autonomic components named Autonomic Communication Elements (ACE).

An ACE represents an engineering effort aimed at providing a component-model through which heterogeneous basic services can be designed and developed in accordance with requirements drawn from self-* properties: ACE internal behaviour, e.g., which prescribes how the ACE has to react to internal and external events, is described by means of a declarative representation (i.e., the self-model): self-model is defined as a set of Extended Finite State Machines which include rules for modifying them to adapt ACE behaviour to the changes of internal and environmental conditions. By programming the self-model of the ACEs, it is possible to engineer the autonomic behaviour, and evolution of service eco-systems: in fact, as described in Section 4.3 and 4.4, the self-model can determine the self-awareness and self-organization capabilities of eco-system components. More advanced services can be created through collections of ACEs that autonomously organize in order to share locally implemented functions, thus realizing a global service eco-system autonomic behaviour. ACEs organization is supported by a collaborative framework that allows the discovery in a distributed setting, and the setup of collaborative behaviour among ACEs. ACEs do not need to be aware of self-models of the other components; in fact, cooperation among ACEs is achieved only by means of interactions performed according to the exposed functions, implemented by each ACE. Self-models and rules for the self-adaptation of ACE behaviour are considered implementation aspects internal to ACEs. Both the component-model and the framework are included in a software development kit, the ACE Toolkit [3], through which ACEs can be designed, developed and deployed. The Toolkit also contains related plug-and-play libraries for exploiting self-management, self-organization, security, and knowledge network capabilities. It is available as an open source project [10].

The paper is organized as follows. Section 2 describes the main principles and primitives for cooperative autonomic behaviour. Then, Section 3 reports a brief summary of the state-of-art of autonomic technologies and solutions for the development of distributed systems. Section 4 describes the vision and the main concepts of an autonomic bio-inspired toolkit designed and developed in the CASCADAS Project. Section 5 analyses some experimental results and a use case concerning a decentralized server farm, considered part of a complex eco-system. Finally, Section 6 concludes the paper with a brief overview of future research.

2   Primitives for Cooperative Autonomic Behaviour

The development of future service eco-systems has to address many challenges such as the increased complexity of large scale networks and their dynamic nature, resource constraints, heterogeneous architectures, absence or impracticality of centralized control and infrastructure, need for survivability, and seamless resolution of failures. These challenges have been successfully solved by Nature during millions of years of evolution.

This evolution developed many biological systems and processes with intrinsic characteristics such as adaptivity to varying environmental conditions, inherent resiliency to failures, collaborative global intelligence which is larger than superposition of individuals, self-organization and evolvability. Inspired by these characteristics, many researchers are currently developing innovative design paradigms to address future network challenges. The objective of this section is providing a brief survey for a better understanding of the potentials for bio-inspired primitives, and to motivate research community to further explore this topic.

2.1   Importance of Interactions

In large scale evolving systems the focus moves from complex algorithms to interactions among huge amount of simple entities [47]. In fact, these systems must be able to operate within rapidly-changing environments composed of a great amount of (possibly simple) independent elements. The complexity of such highly distributed adaptive systems arises out of the multitudes of combinations of ways in which the functioning of its constituent parts can mutually affect and interfere with one another.

This fact has to be considered in the selection of the model to be adopted for describing the behaviour of the processing elements controlling the components in the system and their communications and, possibly, cooperation needed to achieve the overall objectives of the system and guarantee its governance and sustainability. First of all this model must enable a sequential interactive computation, as the elements in the system must continuously interact with their environment (e.g., consisting of other elements): each processing element must react to events coming from its environment, and process them taking into

account its internal state, representing the "history" of the unit; such computations can determine changes in the internal states, and production of other
events towards the environment (and other elements in it).

Although models developed for Multi-Agent Systems (MAS [44]) could seem
able to fulfil these requirements, models based on lightweight processing components provide a better answer to the overwhelming relevance of interactions and
adaptability in complex evolving systems. Specifically, design principles should
ensure the possibility to explore large solution spaces for complex and dynamic
network and service contexts. Moreover, one can imagine that the market pres-
sure determines a sort of natural selection of networks and services (evolving
as digital organisms) which are rewarded for generating models and state diagrams that support certain use-cases and match critical features as specified by
the designer. In fact, on one side, lightweight components offer simple functions
which can be flexibly composed in more complex services, through the creation
of rich interaction patterns; on the other side, simple internal structure of the
components and the orientation to the provisioning of complex services through
composition enable fine degrees of adaptability, in terms of both self-awareness
and self-organization, in order to promptly react to the changes of internal and
environmental conditions.

2.2   Self-awareness and Emergence of Self-organization

Unpredictable events may cause a system to drift away from the desired state,
so it is necessary that the system is able to configure its parameters, at run
time, to enable the optimal response to any changes. In general an autonomic
system is capable of deciding on the fly which components to update, remove or
aggregate in the new configuration. This requires knowledge about the system's
(local and global) state and configuration, as well as a model of the external
environment. Recursively, self-awareness of a single component could be defined
as the component awareness of being a single entity of a global autonomic system, at any point in time [22]. This implies that the component knows certain
internal/external data, transforming them in knowledge (e.g., with reasoning)
and using that knowledge for controlling its behaviour according to plans (e.g.,
by adopting control-loops mechanisms).

In biological systems self-organization is a process in which pattern at the
global level of a system emerges solely from numerous interactions among the
lower-level components of the system. Moreover, the rules specifying interactions among the system's components are executed using only local information,
without reference to the global pattern [8].

Research in various areas, including also physics and cognitive disciplines, has
described process of emergence in terms of the following steps:

- **differentiation:** a difference is recognized in the system and is amplified by
  some internal dynamic or some environmental influence;
- **transforming feedback or coupling:** a transforming bridge is built across
  the differentiation identified in previous step;

– **self-organization:** the system moves to a new state in which the structural framework of the past is superseded by an emergent structure that responds to the feedback loops established in previous step.

Nature, for instance, teaches us that control loops can be combined in emergent algorithms in several ways:

– in the termites algorithm when the threshold of one control loop has been met, then the algorithm changes state to use another control loop responding to a different threshold;
– in the flocking and moth algorithms the meeting of one threshold then leads to another threshold.

These combinations determine emergence of complex social behaviours.

2.3   Bio-inspired Cooperation Primitives

An Autonomic eco-system can be seen as ecology of components capable of abstracting both resources and services; even pieces of data can be abstracted, in principle, in terms of autonomic components. The novelty of this approach is that such components should be able to dynamically interact with each other and self-organize their activities to serve in an adaptive and goal-oriented way the dynamic needs of applications.

This section describes some bio-inspired communication mechanism and protocols, such as gossip, random choice, fields, and gradients that could be used for supporting the cooperation of the components [1].

Gossiping. Gossiping, also known as epidemic communication, is aiming at obtaining an agreement about a certain value of some parameter. Each component broadcasts its belief of value of the parameter to its neighbours, and computation is performed by each component combining the values that it receives from its neighbours. Then the new value is re-broadcasted. The process concludes when there are no further broadcasts.

Random Choice. An example of application of random choice is for establish local identity of large number of components: each component selects a random number to identify itself and communicate it to its neighbours. If the number of choices is large enough, then it is unlikely that components select the same number. Random choice can be combined with gossip to elect super-peers: every component select a value, then gossips to find the minimum. The component with the minimum value becomes the super peer.

Fields. Every component of a certain pattern is considered as a value of field over the discrete space occupied by that pattern. If the density of components is large enough this field of values may be thought of as an approximation of a field on the continuous space.

Gradients. Gradients represent an important primitive in amorphous computing: it implies the estimation of the distance from each component to the nearest component designated as a source. The gradient primitive takes inspiration from the chemical-gradient diffusion process that is crucial to biological development.

2.4   Synchronization in Autonomic Systems

Autonomic systems' behaviours take inspiration from the metaphor of animals' autonomic nervous system. It is well known, from this perspective, that adaptation is the heart of nervous system evolution: a nervous system coming from millions of years of evolution is able to perform complex decision making processes assuring animals' adaptability and survivability. Whatever will be the level of abstraction one may want to consider modelling a nervous system (e.g., multi-level networks of molecules, organelle, neurons), in any case, it seems there are some basic hard-wired decision making and interacting rules embedded into said abstraction elements, at the different levels. Stability and integrity are probably assured by synchronization.

As a matter of fact, S. Strogatz wrote in his book titled Sync: "For reasons we don't yet understand, the tendency to synchronize is one of the most pervasive drives in the universe, extending from atoms to animals, from people to planets" [42]. Perhaps the most celebrated model for synchronization is the Kuramoto model [41], a system of structured ordinary differential equations, which has been used to explain how synchronization is achieved in many engineering, physical and biological systems. Recently, synchronization has also become a subject of study in dynamical systems and control, in the framework of achieving leaderless coordination.

From this perspective, synchronization seems to lie also behind stability and integrity of autonomic systems. Synchronization implies direct or indirect exchange of information, even through spatial and environmental abstractions, and the way this information is used locally; it should be noted, in general, the presence of communication and interaction delays between the components of an autonomic system makes the problem of achieving synchrony even more difficult.

Some traditional paradigms based on approaches such as message-passing, client-server, distributed shared memory, well known in literature, appear to be limited when dealing highly distributed autonomic systems. Spatial and environmental abstractions as a means to drive components interactions and synchronization represent alternative interesting solutions: for example spatial concepts can be realized [51] by means of self-organizing and self-adapting overlay data structures that provide components with context information suitable for driving and coordinating their activities.

As mentioned, the design of hierarchical decision making in autonomic system should also take account of how bio-inspired control loops (after exchanging some information) can be combined into emergent global behaviours [7], for instance, by means of bio-inspired cooperation primitives previous introduced: for instance, in the termites algorithm when the threshold of one control loop has been met, then the algorithm changes state to use another control loop responding to a different threshold; in the flocking and moth algorithms the meeting of one threshold then leads to another threshold. On the other hand, even if all component control loops will be made explicit and the operating regions will be well-defined, interactions between said control loops can result in complex and difficult to understand.

Some design principles can be introduced in order to avoid or, at least, reduce the risks of instability in engineered autonomic systems. Examples are to avoid that multiple independent control loops control the same resources. Moreover, additional autonomic elements could be introduced in order to supervise the evolution of (sub)sets of cooperating independent control loops and enforce on them some changes in policies and parameters in case of critical deviations from planned behaviour [13].

3   State of the Art

There is a wide number of research activities addressing the challenges posed by novel service and future network scenarios, trying to overcome the limitations of current computing/processing and networking/communication frameworks. The state of the art in autonomic computing and service eco-system models and technologies can be rooted mainly to architectures and component models, offering the basic building blocks with which to create applications, services and introducing functions for self-management of services and resources. From a general view, the proposed solutions can be divided among general purpose autonomic systems, multi-agent systems and the autonomic service eco-systems.

3.1   Autonomic Systems

One of the most challenging requirements of future network and service platforms is to adapt to highly dynamic and unpredictable scenarios. This requires the definition and development of autonomic mechanisms to enable self-management (at least for fault, configuration, performance and security management) and self-adaptation to the changes of the execution contexts. With this regard, several prototypical network and service frameworks aimed at enforcing autonomic and adaptable behaviours have been proposed in the past few years. IBM, as part of its autonomic computing initiative [26], has outlined the need for current service providers to enforce adaptability, self-configuration, self-optimization, and self-healing, via service (and server) architectures revolving around feedback loops and advanced adaptation/optimization techniques. Driven by such a vision, a variety of architectural frameworks based on "self-regulating" autonomic components have been recently proposed both by IBM [48] and by independent research centres [11], [19], [31], also with reference to the management of large data centres [49]. The common underlying idea is to couple service components with software components called "autonomic manager" in charge of regulating the functional and non-functional activities of components.

Most of these approaches are typically conceived with centralized or cluster-based server architectures in mind and mostly account for the need of reducing management costs via self-management and self-configuration [50] rather than for the need of providing innovative user-centric services, and definitely not account for the emerging vision of service eco-systems dealing with user-generated content and services.

3.2 Multi-agent Environments

A very similar endeavour - though with a different target scenario - also characterizes several researches in the area of multiagent systems [44]. Multi-Agent Systems researches aim at identifying suitable models and tools to enable the definition of adaptive applications based on autonomous software agents, capable of dynamically interacting with each other and with the computational environment in which they situate, so to achieve global application goals (typically concerned with scenarios related to the orchestration of distributed resources and or with the coordination of organizational activities). Although possibly conceived according to a different architectural model, agents represent de facto sorts of autonomic components which are capable of self-regulating their activities in accord to some specific individual goal and (by cooperating and coordinating with each other) in accord to some global application goal. However, it is worth emphasizing that that Multi Agent Systems does not imply an autonomic behaviour per-se. They are the most diffuse and promising software framework to engineer systems able to show emerging autonomic properties.

At the level of internal structure Belief Desire Intention (BDI) agent systems, as implemented in agent programming systems like Jadex, JACK or Jason [6] or in the context of the Cortex project [5], propose the use of intelligent agents to deal with autonomic and context-aware components. At the core of this model there is a rule-based engine taking actions on the basis of the component internal state that is explicitly represented by means of facts and rules [27], [30].

At the level of multiagent systems and their interactions, agents are mostly supposed to get to know each other via specific agent-discovery services, and are supposed to be able to interact according to "social models", e.g., agent negotiations. This is definitely an important step towards the creation of "agent eco-systems" inspired by some social or economic metaphors. For instance, the ADELFE methodology [4] proposes building complex multiagent systems in accord to the AMAS (Adaptive Multi-Agent System) theory [21]. This focuses on the design of cooperative social interactions between agents, in which each agent possesses the ability of self-organizing its social interactions depending on the individual task it has to solve. However, the identification of models and engineering methods for network and service eco-systems is only in its infancy. More on this will be analyzed later on.

In any case, both most of the proposed approaches for autonomic computing based on autonomic components and most multiagent systems proposal still requires the existence of solid shared middleware substrate to facilitate discovery and interactions between components.

3.3   Autonomic Service Eco-systems

Autonomic service eco-system researches aim at developing models and concepts for autonomic, self-regulating and self-organizing services, capable of co-existing and actively interacting with other services in a service eco-system. In this vision services are seen as the lightweight components which are independent in their existence within a system that are autonomously interacting among each other in a service eco-system in order to achieve higher level goals which are typically more complex services or applications.

In Autonomia framework [18], the autonomic behaviour of the system and the individual applications is handled by the so called mobile agents. The applications and services are composed of software components and system resources and managed by the mobile agents. Each mobile agent is responsible for monitoring a particular behaviour of the system and reacting to the changes accordingly. Mobile agents are seen as the basic enablers of the system autonomic behaviour and are responsible for self-adaptation and self-organization of Autonomia's services and applications.

A slightly different approach is provided by the AutoMate framework [38]. The applications and services in AutoMate are composed out of Autonomic Components which are dynamically composed to applications and services. The components are mapped to the system resources and are controlled by the controller agents. Similar to Autonomia, the autonomic behaviour in the AutoMate framework is handled by the agents and is implemented in form of first order logic rules. The autonomic behaviour is specified at the controller agent level and is distributed over the entire system.

The FOCALE architecture [25] applies a similar concept of autonomic computing in the domain of autonomic network management. For modelling the autonomic behaviour of the network, they propose the idea of mapping business level system constraints down to low level process constraints in an approach called policy continuum [37]. This policy based approach for specifying autonomic system behaviour allows network administrators to specify business level policies for network management (using natural language) like for example different internet connection bandwidth rates for different users, Service Level Agreements (SLA), QoS policies etc. The high level policies, which define the system autonomic behaviour, can be specified by the network designer or administrator using natural language and are translated by the FOCALE framework to the device level policies.

In [46] authors present an Autonomic Element Architecture for developing autonomic systems. They specify the Autonomic System Specification Language (ASSL) framework which allows formal specification and development of autonomic systems. Their architecture is based on the IBM's autonomic system [23], which has been further enhanced and adapted. In the ASSL framework, autonomic system is seen as an entity that consists of the basic building blocks called autonomic elements which are glued together by the Autonomic System interaction Protocol (ASIP). The ASIP is used for all types of interaction among the basic autonomic elements. It supports basic message exchange between the entities as well as the service negotiation functionalities. The ASSL architecture allows expressing autonomic systems as a set of interacting autonomic elements at three main levels: the autonomic system level, the ASIP level, and the autonomic element level. At the autonomic system level it allows specifying system's general characteristics like service level objectives, self-management policies and the autonomic system architecture. At the ASIP level it allows defining all messages that can be exchanged among autonomic elements as well as the channels which can be used. At the autonomic element level, it provides a way for specifying the autonomic element's execution model. As presented in [45], ASSL uses a declarative language for specifying the autonomic behaviors at the different levels of an autonomic system.

In conclusion, looking at the state of the art, none of the analysed approaches seems to address in a holistic way the development and the management of a service eco-system. In particular, none of the above approaches seems to address the problem of globally re-thinking the whole system towards models and architectures specifically conceived for an "ecosystem-based" approach; it uniformly models all the involved components and resources and their interactions according to the bio-inspired principles identified in Section 2. The next section describes how the results achieved in the context of the CASCADAS project overcome the limitations of the current frameworks for autonomic systems.

4   CASCADAS Model and Toolkit

Within the CASCADAS project, a novel agent-based abstraction for the construction of situation-aware and dynamically adaptable services has been developed, known as the Autonomic Communication Element (ACE). ACE abstraction represents an engineering effort aimed at providing a lightweight component-model through which heterogeneous basic services can be designed and developed in accordance with autonomic principles. More sophisticated distributed autonomic systems, up to the scale of complex adaptive service eco-systems, can be built through (self-organized) ensembles of ACEs that share locally available services. ACEs behave on the basis of high-level plans which can dynamically change at runtime, in accordance to the evolution of ACEs' internal state and their local context. ACE component-model is supported by a framework for inter-ACE interactions and cooperation: as detailed in Section 4.1, it adopts self-organising overlay networks interconnecting ACEs and dynamic semantic-based discovery mechanisms to realize nature-inspired structures of components.

In this way the ACE framework differs from other component-based systems which describe context requirements in terms of explicit inter-component dependencies. In fact ACEs do not need to know anything about both dependencies and conflicts, even if, as reported in Section 2.4, stability and integrity should be properly assured during the design phase, for instance, in terms of goal definitions and assignments of interaction rules.

The component-model and the framework were conceived around the consideration of scalability, heterogeneity, portability and reusability aspects. Both are included in the ACE Toolkit through which ACEs can be designed, developed and deployed. The ACE Toolkit has been entirely implemented in Java and is available as open source [10].

4.1   Autonomic Communication Element Model

ACEs component-model and the associated framework enable the concept of service eco-systems. ACEs behave similarly to species individuals in natural eco-systems, where large amounts of heterogeneous entities autonomously enter and leave the environment, interact dynamically among each other, adapting their behaviour and interactions to the way their context evolves. ACEs can autonomously enter, execute in, and leave the ACE execution environment. They can self-adapt their execution to reach an operative state, and dynamically self-organize and self-reorganize among each other without any external human intervention.

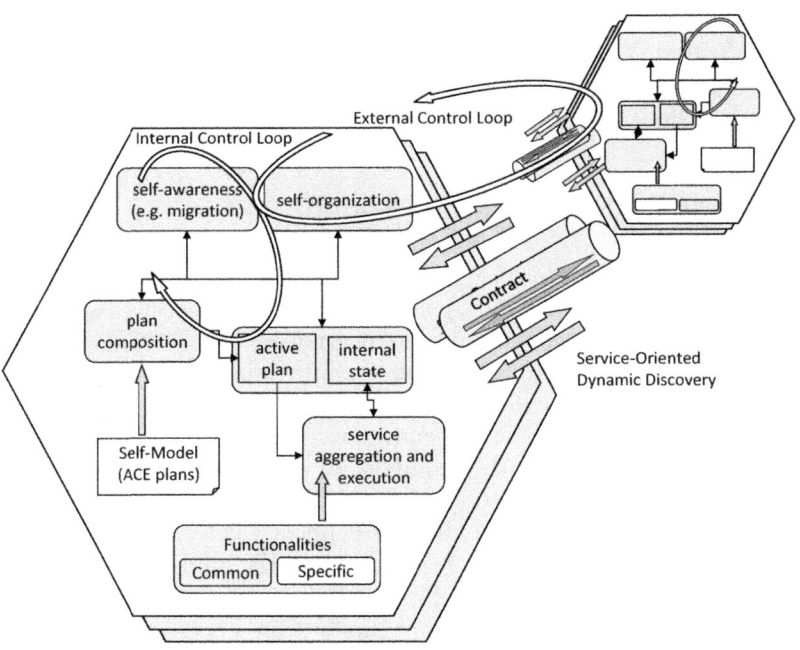

Fig. 1. Autonomic Communication Element model

As presented in Figure 1, ACE model is based on two complementary mechanisms relying on the control loop paradigm. The internal control-loop exploits the self-awareness encompassing reflection, self-control, planning and reasoning. The external control-loop achieves self-organization features for enabling the emergence of global properties and behaviour through very simple cooperative interactions (inspired to biological, social metaphors), and the self-adaptation to the context. Both control loops are governed by the active plan, which describes how the ACE has to react to internal and external events. The active plan determines the locally implemented functions which must be invoked and executed in order to react to the events. The execution of the active plan can be conditioned by the current values of some internal variables describing the ACE internal state. Moreover, the execution of the control loops can change the active plan, according to the modification rules included in the self-model. In order to tackle these challenges and requirements, the ACE model was defined by considering the following design principles.

Explicit Behaviour Definition. When programming an autonomic element, the developer is typically required to provide the high-level policies that define the element's original behaviour [39]. Within the ACE model, such policies are specified through a number of states, along with the transitions that lead the ACE execution process from one state to another, in a way similar to Finite State Machines (FSMs). In terms of ACEs, these policies are called plans. They represent structured behavioural directions an ACE is required to follow, and the overall ACE behaviour can be specified within one or more plans that can execute concurrently or sequentially. The totality of plans specifies the overall behaviour for an ACE, and is enclosed in the ACE's self-model. The current behaviour of an ACE is defined by a subset of these FSMs, which form the active plan governing the control loops.

In terms of plans it can be distinguished between the ACE's "regular" behaviour, which is its behaviour when no events undermining the ordinary execution occur, and the "special cases" which might occur during the plan execution process and which could affect the regular ACE execution process. If such occurrences are foreseen, the ACE behaviour can be enhanced with modification rules, through which it is possible to specify the circumstances under which the original behaviour can be relinquished, along with the new behavioural directions to follow. As detailed in the following section, these modification rules specify the internal or external events that can determine either the activation/deactivation of some plans or the modification of some transactions in the active plan. The ACE behaviour in terms of self-model plans is specified explicitly, with the help of a definition language.

There is a possibility to isolate part of the behaviour specification within the self-model of one ACE and share it as a sort of behavioural pattern among ACEs of similar nature. This "Behavioural Reusability" is achieved through Common Behaviours, i.e., a self-contained set of plans, states and transitions that are made available among a set of ACEs requiring their usage. This feature enable the reuse of already existing plans, so as to ease the development of self-models. However, their use also opens a much wider range of possibilities in terms of dynamic adaptation and extensibility. For instance, consider an ACE-based eco-system where ACE instances (phenotypes) behave according to their self-models (genotypes).

Modular Service Definition and Proactive Service Composition. Within a service eco-system, ACEs find their reason of existence in the provision of functionalities. Functionalities can be regarded as the smallest unit for service provision and are typically concerned with basic capabilities. Functionalities can be combined with other functionalities to form services that are defined in our context as an aggregated set of functionalities. Services can be combined with other functionalities and/or services in order to enlarge their dynamism, complexity and overall usefulness. Services are built proactively by ACEs, in a way specified within the self-model, on the basis of formerly discovered services and existing local functionalities. Once a service has been created, the ACE becomes the owner of the new service and may offer it to other ACEs, according to the conditions defined in the self-model.

Functionalities are defined, by the programmer, at design time in an explicit semantic way similar to the one used for the self-model. Using an explicit definition implies a clear separation from an actual implementation: the functionality definition acts as an interface between the self-model, which is the entity that uses the functionality, and the application code, which implements the functionality.

Service-Oriented Dynamic Discovery. ACEs need to be proactive in finding the right configuration when they are injected in a system. In the vision of service eco-systems, sophisticated services are built through combination of basic services [36], and therefore depend on a mechanism that quickly queries the presence of required services. This is achieved by equipping ACEs with a service discovery mechanism through which ACEs can scout for other ACEs that are offering required services. The service discovery is completely autonomous and self-organized which, in particular, means that every ACE knows which services it offers, as well as which other services it requires in order to execute. Service discovery itself is realized through a protocol named Goal Needed/Goal Achievable (GN/GA).

Stipulated Group Communication. Group communication is fundamental in the interaction model of elements that aim at enabling the design of truly distributed systems. The motivations that might bring two or more ACEs to communicate are heterogeneous in nature, and interacting ACEs might be totally unknown to each other. Therefore, the communication should ideally provide some model for explicit agreement and fast identification of roles within the communication logic itself.

The design of group communication mechanism took inspiration from the use of SLAs; while on the one hand these provide a mean for specifying the terms and conditions without space for ambiguities, on the other hand they hardly suit the needs of lightweight components such as ACEs that, when deployed on mobile devices, might communicate on a non-regular basis. The result of this influence is a contract-based communication that foresees the establishment of a communication contract between the interacting parties. Contracts are essentially channels, which can be bilateral, when created between two ACEs, or multilateral, when created for a larger group of ACEs. Contract establishment needs explicit acceptance from all parties and, once established, contracts provide bidirectional channels through which connected parties can reliably exchange messages. The message exchange can also be "exclusive" within a multilateral contract in that it might target specific destinations only within a bigger group. The use of contract-based communication enables the fast identification of involved parties mentioned above by enforcing a role-based paradigm through which each of the parties is assigned a role. This, in turn, facilitates advanced optional features such as, for instance, contract supervision and provision of security features.

Mobility and Replication. Mobility answers to eventual requests for higher levels of dynamism and long-term optimization ACEs might be needed in specific situations. Consider, for instance, a scenario where an ACE running on a mobile device requires to process large amount of data which will obviously create a significant increase in required workload and will cause suboptimal service provisioning. For that reason ACE can autonomously decide, as specified in the self-model, to move to another physical location from where the service can be appropriately provided.

The mobility and replication features seems especially useful when ACE-based service eco-systems are deployed on a pool of computing resources and promises to increase performance and optimize the resource usage. The ACE framework supports mobility and replication through ACE migration and cloning. The process for the former operation is transparent to the current communication commitments, whereas contracts and connections to other ACEs are moved alongside with the ACE in a seamless way.

4.2   ACE Functional Architecture

From the very general idea of ACEs as autonomic services living in an eco-system, ACE functional architecture has been designed as a biologically inspired computing entity; ACE internal behaviour described in (Figure 1) has been structured and modelled as a set of interoperating modules called organs (Figure 2): each organ is responsible for a particular aspect, providing a specific vital functionality and capable of adapting its own behaviour to the general conditions, and all organs together form a self-contained entity able to interact with its environment. An ACE contains following organs: Facilitator, Executor, Functionality Repository, Manager, Gateway, and Supervision organ.

In a simplified view, ACE consists of the common and the specific part. Common part is available in each and every ACE: it consists of the organs, and a set of common functionalities which are essential to every ACE, such as the service discovery. The specific part on the other hand is ACE specific and varies from ACE to ACE: it consists of the self-model which contains the ACE behaviour specification and a set of ACE-specific functionalities. Therefore, from the programming point of view, ACE developers are required to provide only the specific part.

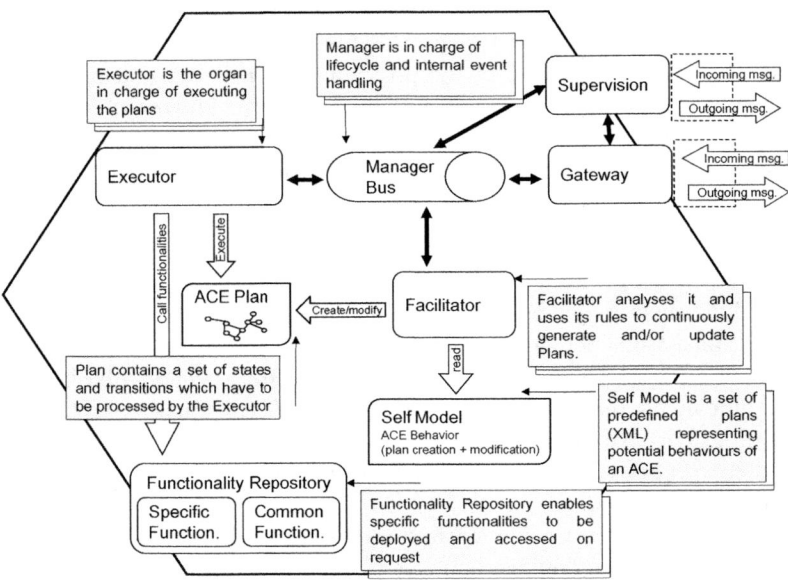

Fig. 2. ACE behaviour modelled as a set of interoperating organs

Facilitator organ is responsible for combining and adapting ACE plans on the basis of the self-model specification, according to internal and external events and context information, to create the active plans suitable to pursue the current goals of the ACE. Plans are forwarded to the Executor for realization. **Executor organ** takes care of executing the active plans which have been provided by the Facilitator. It evaluates the transition conditions and executes the associated actions by requesting, if needed, the Functionality Repository. The Executor processes all the EFSM in the active plan by executing the transactions activated by the internal and external events. **Functionality Repository** is responsible for maintaining ACE functionalities, and for invoking them when requested by the Executor. The executions can access and update the ACE internal state. **Gateway organ** is in charge of the communications with the external world, i.e., with any other ACE. The Gateway provides two communication mechanisms: event-based session-less communications (e.g., for ACE discovery), point-to-point session-capable communications (e.g., for invoking ACE services). The internal structure of ACEs includes some organs in charge of their management; in particular **Manager organ** is responsible for handling the execution of the internal ACE communication, detecting the internal and external events, and controlling the component lifecycle; moreover, **Supervision organ** performs the monitoring of ACE behaviour and the issuing of corrective measures upon detection of hazardous situations: in order to solve situations which can not be locally managed, it can interact, via the gateway, with external supervision services [12]. A detailed description of ACE architecture can be found in [32].

The ACE architecture has some similarities with the IBM's autonomic element architecture [22], but the ACE model can also provide embedded self-management features, implemented by means of Facilitator organ and self-model. Moreover, in ACEs the entire knowledge is represented by the self-model, and the global knowledge emerges out the interactions among the individual ACEs. Furthermore, ACEs provide an easy and effective way for developing autonomic systems, because the developers do not need to specify the individual behaviors for different levels of an autonomic system, as required, for instance, in ASSL framework [46]. In this way, an ACE-based system is highly scalable since parts of the autonomic system can be easily extended reproducing particular ACEs instead of reproducing the entire system. Moreover, it is modular, as the introduction of additional functions does not require adapting the entire autonomic system, but it can be achieved by designing new ACEs. Finally, the approach is truly bio-inspired: autonomic application structures emerge on demand and evolve autonomously in accordance to the behaviors of the individual ACEs.

4.3   Self-awareness in ACE Model

Self-awareness features of ACEs aim at controlling their internal state and events, detecting as well as forecasting in advance critical and/or unplanned situations, and triggering a decision process to select and actuate the corrective actions. In terms of ACEs, self-awareness is seen as an integral part of the ACE autonomic behaviour. Its basic conditions are specified within the self-model and are carried out by the Facilitator organ during the entire ACE lifetime.

Self-model has to specify the way the system behaves, both under the normal and the special conditions, and to define the actions to be performed by the system to autonomously operate in an environment [17], [39]. The autonomic behaviour defined by self-model consists of two integral parts: one that defines the "regular" ACE behaviour and another that defines how this "regular" behaviour needs to be modified when special conditions occurred.

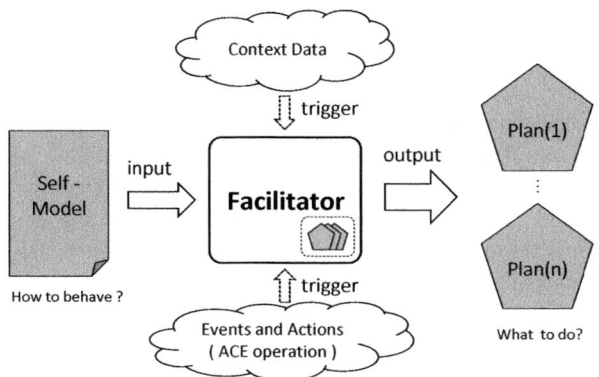

Fig. 3. Facilitator organ: self-model processing and creation/modification of plan(s)

During the ACE initialization phase, the Facilitator loads and analyses the self-model, and creates the initial plan (Figure 3). From there on, it continuously evaluates the need for further plan creations and/or modifications. Facilitator maintains an exact copy of all active plans and continuously observes and evaluates their execution process based on specifications provided within the self-model. Based on the outcome of the evaluation process it creates new and/or modifies the existing plans. A plan specifies the sequence of actions which has to be taken in order to achieve certain goal. It contains states and transitions where a state defines the current status of the plan execution process and transition defines the action to be performed. The new plans and the modified ones are forwarded to the Executor for their actuation. Figure 4 shows the relation between the self-model and the ACE plan.

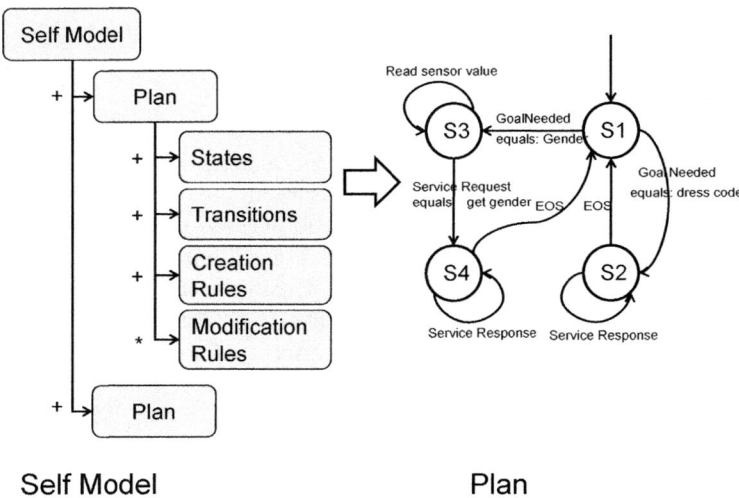

Fig. 4. ACE Self-model and plan

Specifying behavioural directions of an ACE is a task left to the human programmer, so that every aspect can be covered within an arbitrary level of accuracy on the basis of the priorities considered by the programmer. From the implementation point of view, the self-model is defined as an XML structure (Figure 5), based formal language which has been purposely created for this task, and has to be provided to all the ACEs either as a file or a DOM object.

The root element opens one or more plans which in turn are composed by a block of states and transitions each of them carrying the entire set of states and transitions which might be used for this plan. Along with these, each plan also specifies the way the states and transitions need to be assembled into a state machine (*CreationRuleML*) and eventually modified (*ModificationRuleML*) when special conditions occur. Plan creation and modification rules follow the standard RuleML syntax and are defined trough a specification of a number of rules.

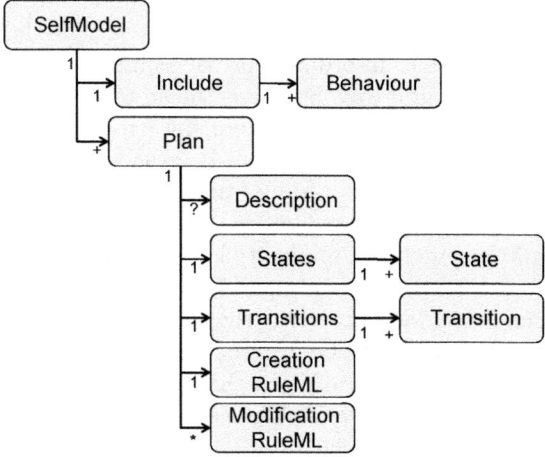

Fig. 5. Definition of the self-model structure

As presented in Figure 6, *states* have univocal identifiers and are defined within plans though a unique ID, friendly name and the desirability level. The friendly name aims at describing the state with a meaningful and user understandable name while the desirability level helps assessing the stability of the execution process. A low desirability level identifies a state the ACE would have probably wanted to avoid within the sphere of ordinary execution but which nevertheless can be entered. The recent state desirability level is continuously monitored by the ACE.

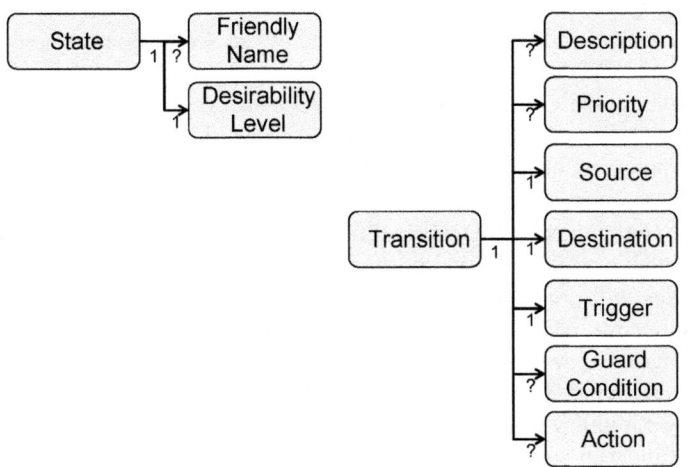

Fig. 6. Self-model structure: states and transitions

Transitions are defined by a number of elements, aimed at univocally defining terms and conditions for the transition to take place. The description is again dedicated to the programmer in order to provide him the possibility to describe the transition in a meaningful way. Source is the state from which the transition will start and destination is the state where the execution process will end as the result of successful transition execution. Priority allows developer to specify the order in which the transitions leading from the same state will be evaluated. Priorities do not add any significant feature to the definition, but allow handling complex cases while keeping the complexity low. A transition provides two levels of filtering. The first level filtering is implied by specifying the transition trigger which is usually an event. More exactly, trigger should specify the type of event upon which the transition should be eventually triggered. Trigger could be also set to "*@auto*" which defines a transition that should be triggered immediately without requiring arrival of an event. The second level filtering can be optionally specified through the guarding conditions. This allows developers to cover cases where more than one transition can take place from the same source state while triggered by the same type of event. For example, let us consider a remote service invocation process resulting in a "service response event". The data returned in the response, but also the information about the success or failure, or the completeness/incompleteness of the executed service, can trigger different transitions and invoke appropriate actions. Even though the first level filtering allow multiple transitions to be triggered, because all of them might satisfy the condition "service response event", the second level filtering allows invoking different transitions, and taking different service execution paths, by using guarding conditions. Guarding conditions allows checking events and their parameter values using first order logic expressions. When an incoming event satisfies both level of filtering, the action defined within the transition will be executed and the plan execution process will move towards the transition destination state. The action typically specifies the local functionality to be invoked.

Creation and modification rules specify how states and transitions should be created and modified within a plan. States and transitions for the "regular" ACE behaviour are created within the block of creation rules, while the ones concerned with "special cases" are enclosed in the block of modification rules.

The following example aims at illustrating the plan creation and modification: ACE-A requires (among others) a functionality F in order to provide its service. For that purpose it needs to find an ACE that provides the functionality F and establish a contract to it in order to use its functionality. For that the plan creation rules will create five transitions in which a goal needed (GN) message is sent, asking for ACEs providing F (`trans.1`), the incoming goal achievable (GA) messages are evaluated and the appropriate ACE is chosen (`trans.2`), the contract is established to the F provider ACE (let us call it ACE-B) (`trans.3`), and the functionality F is used (`trans.4` and `trans.5`). As a result, the five transitions and the corresponding states which describe the "regular" ACE behaviour will be created within a plan as depicted in Figure 7. Let us assume that the ACE-A requires the functionality F to be executed within a predefined time delay. In case the ACE-B cannot meet the defined time constraints, another ACE should be contracted. For this purpose a set of modification rules needs to be specified in which the ACE-B time delay regarding functionality F is monitored. As the result the `trans.4` will be removed and a new transition (`trans.6`) will be created, which will stop the functionality F usage process and will initiate the service discovery. Once a new ACE (e.g., ACE-C) providing the functionality F is found, the `trans.6` is removed and the `trans.4` created, and the ACE-A will continue with its "regular" plan execution, this time using the functionality F provided by the ACE-C.

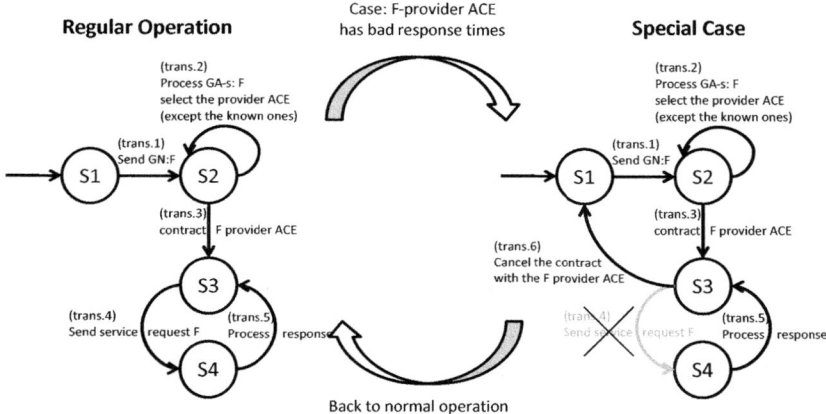

Fig. 7. ACE self-adaptation through plan modification

The previous example illustrates a very simple scenario of ACE self-awareness in a service eco-system, but much more complex ones are possible. In CAS-CADAS framework, an ACE can be aware of any information that is either being monitored by the ACE itself (e.g., execution time delays, broken contracts that are caused by loosing communication to another ACEs, the outcome of the local functionality invocation process), or that is provided by other ACEs (e.g., any type of contextual information, other ACE's execution status).

From the contextual information point of view the CASCADAS framework provides predefined ACEs for monitoring the execution environment conditions, such as the information about the CPU and the memory usage and the number of ACEs executing in a particular environment. This allows ACEs to monitor the execution environment conditions, in particular those of the underlying hardware, and to autonomously decide, in accordance to their self-model, if the current environment is suitable for their execution, or if another "better" environment should be found where the ACE should move to. Therefore, ACEs can be both environment- and self-aware: it is up to the ACE developer to decide how much environment-awareness and how much self-awareness an ACE requires and which conditions must be considered.

In terms of ACEs, self-awareness means also the awareness of the quality/stability of the ACE execution [12], [14]. In CASCADAS framework ACEs have the ability to analyze and introspect their execution process. For this purpose a state desirability level has been introduced (see Figure 6) which helps assessing the stability of the execution process. As mentioned previously, a low desirability level identifies a state the ACE would have probably wanted to avoid within the sphere of ordinary execution but which nevertheless can be entered. Within the ACE self-model, developers are allowed to specify the state desirability level of every state and the actions which should be taken if the service execution process enters an undesirable state. These actions can vary from modifying the currently active plan(s) up to restarting the entire execution process.

The formalism, based on extended FSMs, adopted for defining ACE autonomic behaviour is more advanced than a sequence of actions which was common in earlier autonomic systems [28]. This abstraction has been chosen because it is well-known, easy to understand, and offers the needed descriptive power. Moreover, specifying the autonomic behavior using the extended FSM is more advantageous with respect to other solutions, such as the declarative language approach adopted in ASSL [46]: in fact, it is easier to modify at runtime the FSM-specified autonomic behavior by simply removing the existing and adding new states and transitions, while it is more complicated to re-write at runtime the statements of a specification written by using a declarative language.

Furthermore a clear decoupling of the ACE autonomic behaviour from the business logic of services (e.g., defined in the functionalities stored in the repository) allows more advanced concepts for self-awareness and self-adaptation. This enables the introspection of the service execution process as explained previously, and, at the same time, enables advanced concepts for "forecasting unwanted and/or critical situations". In fact, by considering the state desirability levels, the current plan execution state and the transitions path which have been entered, an ACE is able to forecast if the service execution will progress as wanted, or if actions need to be taken in order to prevent the execution process ending-up in an undesirable and unwanted state. These advanced forecasting concepts, which allow preventing unwanted situations to occur, before they happen, are a very important part of the self-awareness of an autonomic component.

The ACE Toolkit [10] includes a set of common behaviours which provide some of the previously described self-awareness and environment-awareness capabilities that can be used by the developers directly when creating ACEs.

4.4   Self-organization in ACE Model

ACE's self-organization features are in charge of creating and maintaining clusters/ aggregations of ACEs, and self-adapting their behaviour according to the evolutions in their "environment", e.g., by processing events and messages sent by other ACEs. Self-organization capabilities are part of the ACE autonomic behaviour, and are defined within the self-model. In particular, plans for defining self-organization behaviour include transitions which have to cope with events received from external entities though the gateway and the supervision organs.

Self-organization features have a paramount relevance in implementing "modular service definition" and "proactive service composition" principles adopted by ACE framework. According to these principles services are provided by large compositions of relatively simple service features provided by ACEs, which interact for creating composed services or organizing distributed algorithms. The functional composition of ACEs is done by using clustering protocols. In particular, CASCADAS defined and adopted the novel GN/GA protocol which provides a decentralized mechanism for service-oriented dynamic discovery. Its basic flow consists of the following steps (Figure 8):

- **Goal Needed (GN):** this message is sent by an ACE to look for services offered by other ACEs; it contains a (semantic) description of the functionalities the ACE needs to achieve its goals;
- **Goal Achievable (GA):** this message is sent by an ACE to state the task(s) it is able to provide; it transports also a semantic task/service description.

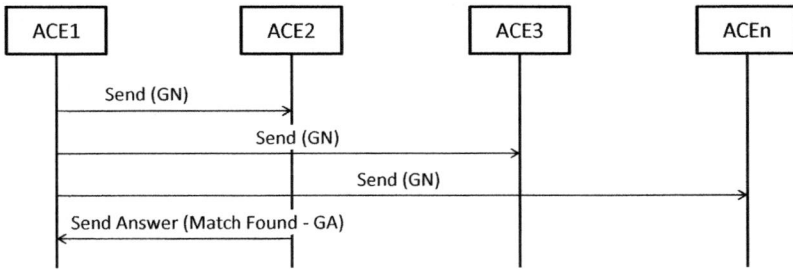

Fig. 8. Flow of the GN/GA protocol

In a distributed setting, the GN/GA protocol is supported by one or more registries where ACEs register in order to join a service discovery overlay network (implemented by the underlying communication middleware, such as REDS), through which the protocol messages are forwarded. These registers manage and provide only the information about the existence of an ACE, but the entire discovery process, i.e., the handling of GN/GA messages, is left to the ACEs, according to the behaviour specified in the self-model (as exemplified in Section 4.3). For instance, in order to scale properly, ACEs must avoid flooding GN requests. This is achieved, for instance, by applying some locality rules and propagating the messages through the discovery overlay.

In order to establish durable cooperation relation between ACEs involved in a service composition, their self-models must plan set-up of contracts. For example, if an ACE A wants to establish a connection to an ACE providing a needed service, e.g., the access to a repository of images, it initially has to find ACEs that are able to offer the required service, e.g., by sending a GN message. When one or more ACEs able to fulfil the required task returns a GA message, ACE A selects one of them, and creates a contract to it.

Although the details of the communication model can vary, the "maximum distance" over which two ACEs can communicate effectively is assumed to be small compared with the size of the entire system and each ACE is usually meant to communicate only with a few neighbours. In order to achieve these requirements, interactions among ACEs, in general offering similar or related services, are usually performed through overlays, whose links are set-up through discovery mechanisms, e.g., GN/GA protocol, and maintained/optimized by means of self-organization algorithms [16], [40].

For instance in *passive clustering* algorithm (1) each of the ACEs interconnected through an overlay may randomly decide to assume the role of matchmaker and initiate a procedure; (2) the match-maker randomly selects two of its own neighbours in the overlay and, if they have the same type, instructs them to link together; (3) if the two chosen nodes were not already directly connected a new link is established between them, and (4) if the number of links must be constant, the match-maker deletes the link with one of its two selected neighbours. An alternative algorithm, named *on-demand clustering*, distinguishes between the initiator of a rewiring procedure and the match-maker; when an ACE randomly decides to assume the role of initiator, it selects one of its neighbours to act as the match-maker, which attempts to connect one of its neighbours to the initiator. Please note that appropriate plans able to cope with clustering algorithms must be included in the self-models of the ACEs which have to participate in an overlay.

Variants of self-organization algorithms are the self-differentiation algorithms that can be used to define decentralized decision algorithms. Self-differentiation algorithms rely on the interactions through an overlay, in order to process and, possibly, modify, the state of two neighbours. Analogously to self-organization algorithms, each ACE in the overlay may randomly assume the role of "matchmaker" and initiate the self-differentiation procedure: it randomly selects two of its neighbours, process their states, and instruct how to change them; in some cases, the match-maker may select only one neighbour, process its state with the one of the selected neighbour, and provide instruction on how to modify them. Also in this case, the logic to deal with self-differentiation is defined in the self-model, possibly with the support of functionalities in the repository. Moreover, the changes of state can determine the creation of a new plan or the modification of a current one performed by the Facilitator, according to the rules in the ACEs' self-models.

Another example of self-organization capability is the one related to the dissemination/exchange of information among ACEs in order to create a global knowledge from local information, by means of gossiping protocols [24]. According to gossiping approach, at each protocol step, an ACE passes to (a small subset of) its neighbours a small amount of data. Algorithms based on gossiping are inspired by the interactions in biological systems, such as those related to the spread of viruses and diseases or the synchronization of blinking of fireflies. Self-aggregation algorithms are an example of these algorithms: during each step of these algorithms, an ACE combines the data received from its neighbours during the previous step, with its local state, updates its state and forwards the combined information to (some of) its neighbours. It was theoretically proven, and confirmed through simulations, that these algorithms converge [29].

By means of self-organization capabilities, systems structured as an ensemble of distributed ACEs are naturally scalable, resilient to faults, and adaptable to changes in environment. They can implement, for instance, decentralized decision algorithms, in case a global "centralized" control is not feasible and global stable information is available: through self-differentiation algorithms and gossiping protocols instead, a global behaviour can emerge from cooperating ACEs. An example of their usage is in [20]: fully decentralized supervision algorithms for fault recovery, load balancing and power savings are implemented as a distributed ensembles of ACEs, cooperating means of gossiping protocols, and self-differentiation algorithms; quasi-optimal supervision behaviour emerged from the local supervision logic executed by single ACEs and the local interaction among them.

A very interesting feature of CASCADAS is the primary role that contextual data have in the overall framework [2]. In the same way as service components self-organize with each other, the pieces of the contextual information similarly self-compose. By the way, this mechanism enables the creation of knowledge networks and cognitive capabilities [33].

These self-organization capabilities must be included in the autonomic behaviour of each of the ACEs, by means of plans and rules in their self-models, with the possible support, if required, of more complex logic in their functionality repositories. Some parts of this behaviour could be defined as Common Behaviour shared among most of the ACEs. Examples are the behaviour to deal with GN/GA protocol or that for clustering algorithms.

4.5   CASCADAS ACE Toolkit Experimental Validation

Performance of the ACE Toolkit has been evaluated from the perspective of a distributed system. For this purpose several experiments have been performed aiming at measuring: threads, memory usage, and communication delay of ACE applications. Evaluating the ACE Toolkit performance and resource consumption from the distributed system perspective was important because ACEs are created to operate in such environments.

For example a set of experiments analyzed the memory consumption of ACE based applications. For this experiment, 100 ACEs have been created and sequentially started one after another. Each time a new ACE has been created and started, the memory that is allocated for the application has been measured. Different Java garbage collectors have been utilized and the results have been compared accordingly. Figure 9 depicts the memory allocation patterns when using the garbage collectors ParallelOldGC, ParallelGC, SerialGC whereas Figure 10 presents the memory allocation pattern when using the incremental garbage collector (IncGC) [43].

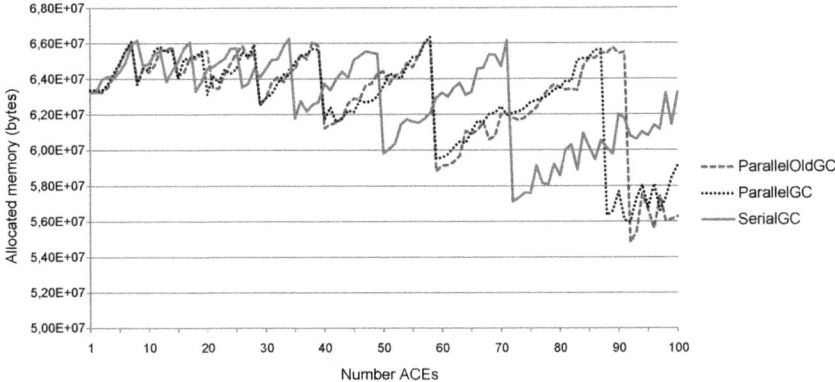

Fig. 9. Memory consumption when using ParallelOldGC, ParallelGC and SerialGC garbage collectors

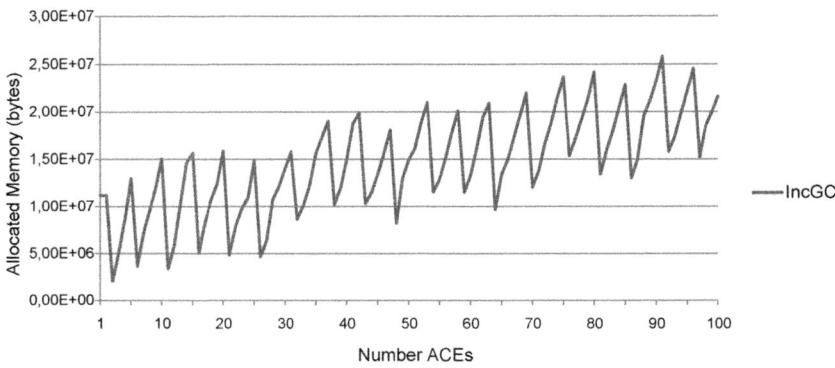

Fig. 10. Memory consumption when using incremental garbage collector IncGC

While the ParallelOldGC, ParallelGC, SerialGC garbage collectors exhibit a similar behaviour suggesting a constant amount of allocated memory, IncGC, shows a memory consumption curve that linearly increases with the amount of ACEs. The oscillatory trend of all garbage collectors is due to the fact that they are not permanently cleaning up the memory but are only running periodically with a certain time delay between each execution. In conclusion, the memory consumption of ACE based applications exploits in an optimal way the characteristics of the garbage collector. In particular, this allows that the amount of allocated memory grows only linearly with the number of ACEs within the application.

In another set of experiments, the delays in inter-ACE communication in a distributed test-bed have been measured. Communication characteristics have been tested experimentally through executions on a test-bed composed by machines distributed over a LAN.

The experiment was performed in the following way: In the initial setting, two ACEs find each other and start communicating through a simple request-response protocol. To this extent, a request message is sent every 500ms, with the source ACE recording the transmission time. Once the message is received, a reply message is sent back to the source where the reception time is recorded to measure the communication round-trip time. Every two minutes, a new ACE joins the system. This is quickly found by other ACEs, via the GN/GA protocol, and suddenly involved in the request-response process. The communication delay is calculated as the difference between the transmission time of the request message and the reception time of the reply message. The timings are recorded at ACE application level and to be more precise, the time delay recorded includes the actual network delay and the time needed for the packet to be processed (at both source and destination). Nevertheless, considering the fact that the message processing at the ACE level means simply sending the same message back to its source, the time required for message processing will impact the recorded timings on a lesser extent.

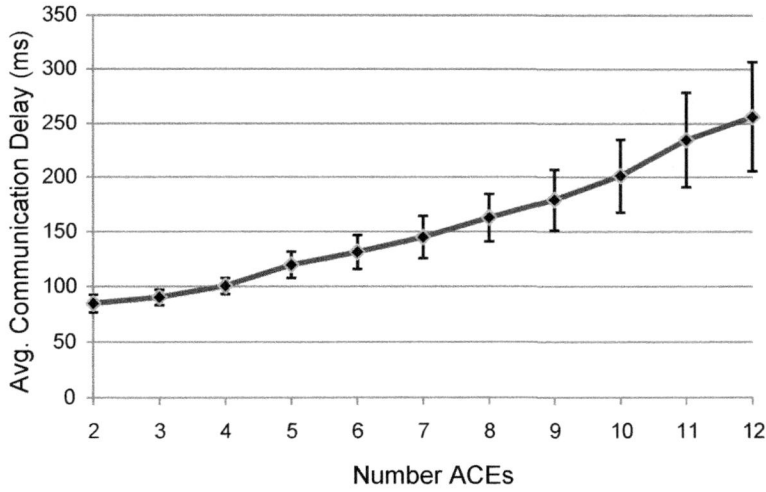

Fig. 11. Average inter-ACE communication delay

As presented in the Figure 11, the communication delay increases linearly with the number of ACEs. The delay variations can be mainly imputed to variations in the message processing times. The increase of communication delays with the number of ACEs shows a linear scalability in distributed ACE applications.

In conclusion of these and other experiments reported in [32], it is possible to see that the ACE architecture is rather scalable along many dimensions (e.g., memory, threads, communication delay) in distributed setting. This supports the applicability of the ACE model in large autonomic communication scenarios.

5   Service Eco-systems

Current trends in Telecommunications and Future Internet push towards the development of service eco-systems capable of enabling the aggregation of virtualized resources and composition of services and functionalities. An eco-system is generally defined as a community of organisms living in a particular environment and the physical elements in that environment with which they interact. Within each eco-system, there are habitats which may also vary in size. A habitat is the place where a population lives. A population is a group of living organisms of the same kind living in the same place at the same time. All of the populations interact and form a community. The community of living things interacts with the non-living world around it to form the eco-system. The habitat must supply the needs of organisms, such as food, water, temperature, oxygen, and minerals. If the population's needs are not met, it will move to a better habitat. Two different populations cannot occupy the same niche at the same time, however. So the processes of competition, predation, cooperation, and symbiosis occur.

Adopting this metaphor, an autonomic eco-system consists of interactive collections of individual autonomic components that are abstracting resource, services and data in order to deliver services to humans and other autonomic components [15], [35]. These autonomic components could manage their internal behaviour and their relationships with other autonomic components according to higher level policies. In other words, autonomic eco-systems are complex systems that comprise many interconnected components operating at different time scales in a largely independent fashion that manage themselves to satisfy high-level system requirements.

Adopting another perspective, also a complex organism (e.g., a person or an animal) can be seen as an eco-system of living organs and cells. Autonomic components could be considered like individual cells (characterized in terms of phenotypes) with a certain genotype. As for living cells, genotype/phenotype correlations are a basic ingredient for enabling dynamic evolution of components being part of said organism (i.e., an eco-system in the adopted metaphor).

Consider, the example of the biological response to stress of said organism: this response implies the maintenance of homeostasis in the presence of certain challenging, stressful (or perceived such) situations. The brain this organism activates many neuronal circuits linking centres for sensory, motor, autonomic, neuro-endocrine, cognitive, and emotional functions, etc. in order to adapt to the situation. So, in this example, we may consider the autonomic behaviour as a global response of the organism, activated by the brain, to adapt to stressful (or perceived such) situations. Neurobiology, for example, says that in those situations some gene receptors are responsible to mediate signals, through certain circuitries, to organs. So, progressing in this neurobiological metaphor, it is important to analyze which are the specific circuitry (phenotypes), signals and related genes (i.e., genotype/phenotype correlations), which are involved in the specific stress response (which is in our example, autonomic modulation). By the way, this aspect might be a first step in understanding how genotype/phenotype correlations relate to autonomic behaviours.

Next section will describe a use-case developed in the ICT Project CAS-CADAS, as an example of feasibility study of an autonomic service eco-system. The use-case concerns a decentralized server farm considered as part of a living organism. It will be shown that the autonomic reaction mediated by distributed components (i.e., cells) determines a response to optimize the global behaviour of the decentralized server farm. Autonomic components behave locally according to specific rules (coded into the self-model, i.e., in the metaphor, its genotype) and communicate with each other to optimize a global behaviour of the server farm (in the metaphor, gene receptors mediate signals to organs of the organism).

5.1   Use Case: Decentralized Server Farm as Part of an Organism

Let us consider a decentralized server farm (as a part of an organism) created by interconnecting computing nodes, including both servers, clusters, and, also, users' devices, through a geographical-wide network (e.g., the Internet), in or-der to form a cloud of computing resources interconnected though an overlay network [34]. According to the supervision framework, each element in the sys-tem under supervision (e.g., each computing resource in the decentralized server farm) is equipped with an ACE in charge of its supervision. All these supervising ACEs share the same behaviour, and, thus, the same self-model, and are inter-connected through a (self-organized) overlay. The supervising ACEs self-adapt their behaviour according to their internal state, the state of the supervised resource, and the interactions of their neighbours in the overlay. The supervis-ing ACEs interwork and cooperate though the overlay, e.g., according to some self-differentiation algorithm, to optimize some global parameters representing the "performance" of the whole system (e.g., reduction of failures, reduction of execution time, reduction of power).

The behaviour of the ACEs is exemplified by considering the definition of power saving algorithm for a decentralized server farm; the objective of the algorithm is to reduce the energy consumed by the whole system, by limiting the impacts of the time for executing tasks. The logic stems from the fact that a node in stand-by consumes much less energy than a node that is idle. Thus, a group of nodes with low utilization is a waste of energy considering that the same work could be executed by a smaller number of nodes, by putting in stand-by the remaining ones. The behaviour of ACEs supervising the computing nodes can be organized in the following states, each of them corresponding to a plan defined in their self-model (whose specification is reported in [9]):

1. **Underused**, i.e., the node supervised by the ACE A has load lower than a given "underused" threshold:
   (a) if A finds a neighbour B in the overlay, which is able to take all load of its supervised node, then A transfers this load to node supervised by B and put the supervised node in stand-by;
   (b) if A gets a request from a neighbour B to take (some of) the load of its supervised node, it checks if it can fulfil the request; in case it can, it takes the load from the node supervised by B and transfer it to its supervised node;

2. **Normal**, i.e., the node supervised by the ACE A has load between "under-used" and "overloaded" thresholds:
   (a) A does not process events concerning power saving;
3. **Overloaded**, i.e., the node supervised by the ACE A has load higher than a given "overloaded" threshold:
   (a) if A finds a neighbour B, which is able to take part of the load of its supervised node, then A transfers it to node supervised by B;
   (b) if (3.a) fails, A selects one of its neighbour which is in stand-by, wakes it up, and transfers part of its load to it.
4. **Stand-by**, i.e., the node supervised by the ACE A is in stand-by:
   (a) when A receives a request from an overloaded neighbour B, A wakes up its supervised node and transfers some of the load from the node supervised by B to its supervised node.

To avoid state oscillation, an ACE has to wait for some time after being woken up before it can execute (1.a). Moreover, to reduce the number of failures in looking for a neighbour to wake-up, an overloaded node has to wait for some time after a failed attempt in waking up a node, before a new execution of (3.b).

The effectiveness of the approach was analyzed through simulations. Figure 12 compares the energy consumed by a system executing only rules implementing load balancing policies (i.e., 1.b, 2.a, 3.a), with that consumed by a system executing also rules for energy savings. Energy savings rules save about 14% of energy, with a limited impact on execution time (about 5% in stable state). During the recovery phase from a traffic peak (simulation cycles 200-300) there is a maximum increment of 45% in the average execution time, due to delay in information propagation across the whole system. During traffic stability periods, the algorithm computes a quasi-optimal solution.

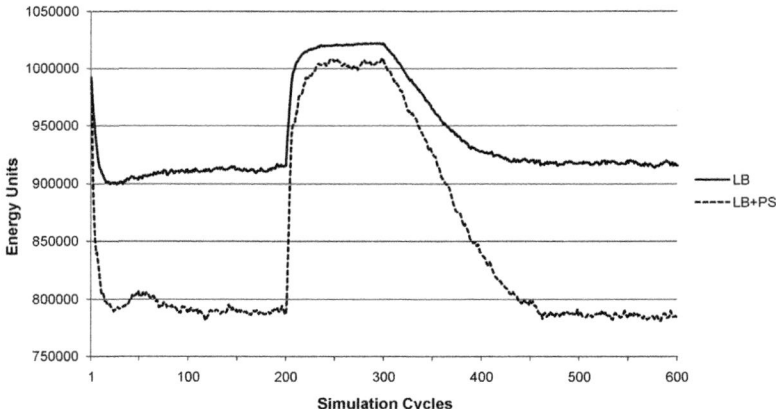

Fig. 12. Comparing energy consumption of a system impl. only rules for load balancing (LB), with that used by a system adopting also rules for energy savings (LB+PS).

6   Conclusions

It is easy imagining a future when millions of cell phones and smart digital devices will embed processing and storage power and communications features. This evolution will create a pervasive and complex computational and networking environment fostering the evolution towards eco-systems of resource, services and data. This will determine a completely new set of challenges and problems to be addressed related to massively distributed processing and storage and to highly complex and dynamic communications networks. Autonomic technologies coupled with bio-inspired principles can provide strong elements for facing such challenges.

This paper has addressed the development of autonomic bio-inspired eco-system: specifically it has described the model of autonomic component implemented and experimentally validated within the EU project CASCADAS. Moreover it has elaborated how self-organization (and other self-* feature) emerges by using bio-inspired interactions and cooperation. In conclusion, the adopted approach has gone beyond a traditional mechanistic one, whereby concepts derived from biology are applied to explain what happens in a complex eco-system: in fact, by programming the self-model of the ACEs, it is possible to engineer the autonomic behaviour, and evolution of service eco-systems. Moreover, ACEs differ from more tradition autonomic elements, as from the terminology of autonomic computing [39], for a main reason: ACE instances in a self-managed eco-system can be considered the phenotypes of living entities (species) whose behaviour and evolution are governed by their genotypes (i.e. the self-models): ACE framework and Toolkit innovation stands mainly in this bio-mimetic approach. In this way, the proposed model has opened the black box of an autonomic bio-inspired eco-system in order to progress the understanding of how it generates its dynamic behaviours. Finally, the paper analysed some experimental results and a use case concerning a decentralized server farm, considered part of a complex eco-system.

Yet, there are many foundational challenges to be addressed to effectively achieve the deployment of autonomic bio-inspired eco-systems such us: meeting a top-down approach (e.g., based on autonomic mechanisms) with a grassroots bottom-up one (e.g., based on much simpler bio-inspired mechanisms), creating and managing global knowledge from data and metadata and the problems of potential phase transitions in largely distributed interacting systems. It is argued that serious progresses in this research avenue could be achieved only through an interdisciplinary approach building on results from diverse fields such as biology, physics, game theory, economics, complexity theory and autonomic computing.

Acknowledgments. The authors would like to acknowledge the European Commission for funding the Integrated Project CASCADAS "Component-ware for Autonomic, Situation-aware Communications, And Dynamically Adaptable Services" (FET Proactive Initiative, IST-2004-2.3.4 Situated and Autonomic Communications) within the 6th IST Framework Program.

References

1. Abelson, H., Beal, J., Sussman, G.J., Abelson, H., Beal, J., Sussman, G.J.: Amorphous computing, MIT Technical Report CSAIL TR 2007 030
2. Baumgarten, M., Bicocchi, N., Kusber, R., Mulvenna, M.D., Zambonelli, F.: Self-organizing knowledge networks for pervasive situation-aware services. In: IEEE International Conference on Systems, Man and Cybernetics (SMC 2007), pp. 1–6. IEEE Computer Society (2007)
3. Benko, B., Brgulja, N., Höfig, E., Kusber, R.: Adaptive services in a distributed environment. In: Proceedings of the Eighth International Workshop on Applications and Services in Wireless Networks (ASWN 2008), pp. 66–75. IEEE Computer Society (2008)
4. Bernon, C., Gleizes, M.-P., Peyruqueou, S., Picard, G.: ADELFE: A Methodology for Adaptive Multi-Agent Systems Engineering. In: Petta, P., Tolksdorf, R., Zambonelli, F. (eds.) ESAW 2002. LNCS (LNAI), vol. 2577, pp. 156–169. Springer, Heidelberg (2003)
5. Biegel, G., Cahill, V.: A framework for developing mobile, context-aware applications. In: Proceedings of the Second IEEE International Conference on Pervasive Computing and Communications (PerCom 2004), pp. 361–365. IEEE Computer Society (2004)
6. Bordini, R.H., Dastani, M., Dix, J., Fallah-Seghrouchni, A.E. (eds.): Multi-Agent Programming: Languages, Platforms and Applications, vol. 15. Springer, Heidelberg (2005)
7. Breza, M., McCann, J.: Can fireflies gossip and flock?: The possibility of combining well-know bio-inspired algorithms to manage multiple global parameters in wireless sensor networks without centralised control. (2008), http://www.doc.ic.ac.uk/~mjb04/structure_of_emergence.pdf (visited on November 25, 2010)
8. Camazine, S., Franks, R.N., Sneyd, J., Bonabeau, E., Deneubourg, J.L., Theraula, G. (eds.): Self-Organization in Biological Systems. Princeton University Press, Princeton (2001)
9. CASCADAS-Project: ACE-based supervision and ACE tool-kit embedded supervision: final architecture, and demonstrators (2008), www.cascadas-project.org
10. CASCADAS-Project: ACE toolkit repository (2009), http://sourceforge.net/projects/acetoolkit/
11. Chen, Y., Iyer, S., Liu, X., Milojicic, D., Sahai, A.: SLA Decomposition: Translating service level objectives to system level thresholds. In: Proceedings of the Fourth International Conference on Autonomic Computing (ICAC 2007). IEEE Computer Society (2007)
12. Deussen, P.H., Baumgarten, M., Mulvenna, M., Manzalini, A., Moiso, C.: Autonomic re-configuration of pervasive supervision services. In: International Conference on Emerging Network Intelligence, pp. 33–38. IEEE Computer Society (2009)
13. Deussen, P.H., Baumgarten, M., Mulvenna, M., Manzalini, A., Moiso, C.: Component-ware for autonomic supervision services - the CASCADAS approach. IEEE International Journal on Advances in Intelligent Systems 3, 87–105 (2010)
14. Deussen, P.H., Höfig, E.: Self-organizing service supervision. In: 2nd International Conference on Bio-Inspired Models of Network, Information and Computing Systems, pp. 245–246. IEEE Computer Society (2007)

15. Deussen, P.H., Höfig, E., Manzalini, A.: An ecological perspective on future service environments. In: Proceedings of the 2008 Second IEEE International Conference on Self-Adaptive and Self-Organizing Systems Workshops, pp. 37–42. IEEE Computer Society (2008)

16. Di Marzo Serugendo, G., Foukia, N., Hassas, S., Karageorgos, A.: Self-Organisation: Paradigms and Applications. In: Di Marzo Serugendo, G., Karageorgos, A., Rana, O.F., Zambonelli, F. (eds.) ESOA 2003. LNCS (LNAI), vol. 2977, pp. 1–19. Springer, Heidelberg (2004)

17. Dobson, S., Denazis, S., Fernández, A., Gaïti, D., Gelenbe, E., Massacci, F., Nixon, P., Saffre, F., Schmidt, N., Zambonelli, F.: A survey of autonomic communications. ACM Transactions on Autonomous and Adaptive Systems 1, 223–259 (2006)

18. Dong, X., Hariri, S., Xue, L., Pavuluri, S., Zhang, M., Rao, S.: Autonomia: an autonomic computing environment. In: IEEE International Conference on Performance, Computing, and Communications, pp. 61–68. IEEE Computer Society (2003)

19. Farha, R., Kim, M.-S., Leon-Garcia, A., Hong, J.W.-K.: Towards an Autonomic Service Architecture. In: Magedanz, T., Madeira, E.R.M., Dini, P. (eds.) IPOM 2005. LNCS, vol. 3751, pp. 58–67. Springer, Heidelberg (2005)

20. Ferrari, L., Manzalini, A., Moiso, C., Deussen, P.H.: Highly distributed supervision for autonomic networks and services. In: Advanced International Conference on Telecommunications, pp. 111–116. IEEE Computer Society (2009)

21. George, J.P., Edmonds, B., Glize, P.: Making self-organizing adaptive multi-agent systems work. In: Bergenti, F., Gleizes, M.P., Zambonelli, F. (eds.) Methodologies and Software Engineering for Agent Systems, pp. 321–340. Springer, Heidelberg (2004)

22. Horn, P.: Autonomic computing: IBM's perspective on the state of information technology (2001), http://www.research.ibm.com/autonomic/manifesto/ (visited on November 25, 2010)

23. IBM-Corporation: An architectural blueprint for autonomic computing (2006), white Paper http://www-01.ibm.com/software/tivoli/autonomic/ (visited on November 25, 2010)

24. Jelasity, M., Montresor, A., Babaoglu, O.: Gossip-based aggregation in large dynamic networks. ACM Transactions on Computer Systems 23, 219–252 (2005)

25. Jennings, B., Van der Meer, S., Balasubramaniam, S., Botvich, D., Foghlu, M.O., Donnelly, W., Strassner, J.: Towards autonomic management of communications networks. IEEE Communications Magazine 45(10), 112–121 (2007)

26. Kephart, J.O., Chess, D.M.: The vision of autonomic computing. IEEE Computer 36, 41–50 (2003)

27. Klein, C., Schmid, R., Leuxner, C., Sitou, W., Spanfelner, B.: A survey of context adaptation in autonomic computing. In: International Conference on Autonomic and Autonomous Systems, pp. 106–111. IEEE Computer Society (2008)

28. Koehler, J., Giblin, C., Gantenbein, D., Hauser, R.: On autonomic computing architectures, IBM Zurich Research Laboratory, Computer Science Tec. Rep. RZ 3487-99302 (2003), www.zurich.ibm.com/pdf/ebizz/idd-ac.pdf (visited on November 25, 2010)

29. Kowalczyk, W., Vlassis, N.: Newscast EM. In: Advances in Neural Information Processing Systems, pp. 713–720. MIT Press (2005)

30. Li, J., Martin, P., Powley, W., Wilson, K., Craddock, C.: A sensor-based approach to symptom recognition for autonomic systems. In: International Conference on Autonomic and Autonomous Systems, pp. 45–50. IEEE Computer Society (2009)

31. Liu, H., Parashar, M., Hariri, S.: A component-based programming model for autonomic applications. In: International Conference on Autonomic Computing, pp. 10–17. IEEE Computer Society (2004)
32. Manzalini, A., Brgulja, N., Minerva, R., Moiso, C.: Specification, development, and verification of CASCADAS autonomic computing and networking toolkit. In: Cong-Vinh, P. (ed.) Formal and Practical Aspects of Autonomic Computing and Networking: Specification, Development and Verification. IGI, Hershey (2010)
33. Manzalini, A., Deussen, P.H., Nechifor, S., Mamei, M., Minerva, R., Moiso, C., Salden, A., Wauters, T., Zambonelli, F.: Self-optimized cognitive network of networks. The Computer Journal (2010), doi: 10.1093/comjnl/bxq032
34. Manzalini, A., Minerva, R., Moiso, C.: Exploiting P2P solutions in telecommunication service delivery platforms. In: Antonopoulos, N., Exarchakos, G., Li, M., Liotta, A. (eds.) Handbook of Research on P2P and Grid Systems for Service-Oriented Computing: Models, Methodologies and Applications, pp. 937–955
35. Manzalini, A., Minerva, R., Moiso, C.: Autonomic clouds of components for self-managed service ecosystems. Journal of Telecommunications Management 3(2), 164–180 (2010)
36. Manzalini, A., Zambonelli, F.: Towards autonomic and situationaware communication services: the CASCADAS vision. In: 1st IEEE Workshop on Distributed Intelligent Systems, pp. 383–388. IEEE Computer Society (2006)
37. Van der Meer, S., Davy, S., Davy, A., Carroll, R., Jennings, B., Strassner, J.: Autonomic networking: Prototype implementation of the policy continuum. In: 1st International Workshop on Broadband Convergence Networks (BcN). IEEE Computer Society (2006)
38. Parashar, M., Liu, H., Li, Z., Matossian, V., Schmidt, C., Zhang, G., Hariri, S.: AutoMate: Enabling autonomic applications on the grid. Cluster Computing 9, 161–174 (2006)
39. Parashar, M., Hariri, S.: Autonomic Computing: An Overview. In: Banâtre, J.-P., Fradet, P., Giavitto, J.-L., Michel, O. (eds.) UPP 2004. LNCS, vol. 3566, pp. 257–269. Springer, Heidelberg (2005)
40. Saffre, F., Tateson, R., Halloy, J., Shackleton, M., Deneubourg, J.L.: Aggregation dynamics in overlay networks and their implications for self-organized distributed applications. The Computer Journal 52, 397–412 (2009)
41. Strogatz, S.H.: From Kuramoto to Crawford: exploring the onset of synchronization in populations of coupled oscillators. Physica D: Nonlinear Phenomena 143, 1–20 (2000)
42. Strogatz, S.H. (ed.): SYNC: The Emerging Science of Spontaneous Order. Hyperion Press (2003)
43. Sun-Microsystems: Tuning garbage collection with the 5.0 java virtual machine docu. (2003), http://java.sun.com/docs/hotspot/gc5.0/gc_tuning_5.html
44. Valckenaers, P., Sauter, J., Sierra, C., Rodriguez-Aguilar, J.A.: Applications and environments for multi-agent systems. Autonomous Agents and Multi-Agent Systems 14, 61–85 (2007)
45. Vassev, E., Paquet, J.: ASSL - Autonomic System Specification Language. In: Proceedings of the 31st IEEE Software Engineering Workshop (SEW 2007), pp. 300–309. IEEE Computer Society (2007)
46. Vassev, E., Paquet, J.: Towards an autonomic element architecture for ASSL. In: Proceedings of the 2007 International Workshop on Software Engineering for Adaptive and Self-Managing Systems (SEAMS 2007), p. 4. IEEE Computer Society (2007)

47. Wegner, P., Arbab, F., Goldinand, D., McBurneyand, P., Luck, M., Robertson, D.: The role of agent interaction in models of computing: Panelist reviews. Electronic Notes in Theoretical Computer Science (ENTCS) 141, 181–198 (2005)
48. White, S.R., Hanson, J.E., Whalley, I., Chess, D.M., Kephart, J.O.: An architectural approach to autonomic computing. In: 1st International Conference on Autonomic Computing, pp. 2–9. IEEE Computer Society (2004)
49. Xu, J., Zhao, M., Fortes, J., Carpenter, R., Yousif, M.: On the use of fuzzy modeling in virtualized data center management. In: 4th International Conference on Autonomic Computing, p. 25. IEEE Computer Society (2007)
50. Zambonelli, F.: Self-management and the many facets of "Nonself". IEEE Intelligent Systems 21, 50–56 (2006)
51. Zambonelli, F., Mamei, M.: Spatial Computing: An Emerging Paradigm for Autonomic Computing and Communication. In: Smirnov, M. (ed.) WAC 2004. LNCS, vol. 3457, pp. 44–57. Springer, Heidelberg (2005)

A Logical Approach to Data-Aware Automated Sequence Generation

Sylvain Hallé[1], Roger Villemaire[2], Omar Cherkaoui[2], and Rudy Deca[2]

[1] Université du Québec à Chicoutimi, Canada
shalle@acm.org
[2] Université du Québec à Montréal, Canada
{villemaire.roger,cherkaoui.omar}@uqam.ca

Abstract. Automated sequence generation can be loosely defined as the algorithmic construction of a sequence of objects satisfying a set of constraints formulated declaratively. A variety of scenarios, ranging from self-configuration of network devices to automated testing of web services, can be described as automated sequence generation problems. In all these scenarios, the sequence of valid objects and their data contents are interdependent. Despite these similarities, most existing solutions for these scenarios consist of *ad hoc*, domain-specific tools. This paper stems from the observation that, when such "data-aware" constraints are expressed using mathematical logic, automated sequence generation becomes a case of satisfiability solving. This approach presents the advantage that, for many logical languages, existing satisfiability solvers can be used off-the-shelf. The paper surveys three logics suitable to express real-world data-aware constraints and discusses the practical implications, with respect to automated sequence generation, of some of their theoretical properties.

1 Introduction

Historically, the advancement of computing has been marked by the development of successive abstractions in the description of the tasks to be accomplished by a system. As an example, in the field of programming, the advent of the assembly language, followed by structured programming languages, allowed users to progressively distance themselves from technical and hardware issues to concentrate on a logical description of the work to be done.

This evolution towards more abstract descriptions continues to this day. Until recently, the development and use of a system followed an approach which could be called *imperative*: although the languages and the formalisms evolved towards more abstract concepts, the basic principle always consisted of describing the tasks to be carried out in order to produce a desired result. However, a number of fields in computing have been transformed in recent years with the advent of a new, *declarative* approach.

In a declarative paradigm, the results, rather than the tasks, are described by means of a language. The actual way in which these results should be obtained

M.L. Gavrilova et al. (Eds.): Trans. on Comput. Sci. XV, LNCS 7050, pp. 192–216, 2012.

is not specified. The new job for a user is no longer to design and express a sequence of steps that the system should follow, but rather to describe as precisely as possible the data or functionalities expected from the system. While the imperative approach "commands", the declarative approach "demands".

This raises a fundamental problem: to develop automated techniques to produce a result that satisfies some declarative specification. The premise of this work is the following: mathematical logic represents an appropriate formal foundation for the resolution of this problem. Indeed, mathematical logic is essentially a declarative language: by means of logical connectives, it is possible to build statements that become true or false depending on the context on which they are interpreted.

First, in Section 2, the paper studies the concept of sequential constraints in computing and shows by a couple of examples their widespread presence in a variety of scenarios, ranging from web service compositions to the management of network devices. Section 2.3 then shows how the automated generation of sequences of operations can be put to numerous uses in the aforementioned scenarios. Depending on the field, automated sequence generation becomes an instance of self-configuration, test case generation, or web service composition problems.

Second, in Section 3, the paper provides a simple formal model for representing operations and sequences of operations carrying simple data payloads such as command parameters, XML message elements, or function arguments. It shows how all the presented scenarios can be appropriately represented under this model, thereby giving a common, domain-independent formal foundation for the study of sequential dependencies in computing.

Finally, the paper advocates the use of logical methods for representing such constraints in a declarative fashion. Automated sequence generation then becomes a mere instance of satisfiability solving for a set of logic formulæ. An interesting consequence is that, for many logics, automated satisfiability solvers are readily available, and can be used as general purpose engines, thus removing the need for *ad hoc*, domain-specific solutions. In addition, well studied properties of these logics, such as undecidability theorems or complexity results, apply directly to the computing scenarios they are used to model.

This formal model reveals an important and often overlooked characteristic of sequential properties. In many cases, the sequence of valid objects and their respective data contents cannot be treated separately. Yet, we shall see that traditional logics are generally ill-equipped for expressing such kinds of constraints.

To the best of the authors' knowledge, this is the first attempt at a domain-independent study of logic-based automated sequence generation. Hence, the paper intends to be a road map for future work on that topic. To this end, a set of three logics suitable for that task are described in Sections 4 to 6. The same sample properties of a real-world scenario are modelled in each of them; tools and techniques for automatically generating sequences satisfying the resulting formulæ are then presented.

2   Sequential Aspects in Computing

The first part of this paper deals with the description of valid sequences of operations in various fields, and describes how such sequences of operations can be
automatically generated. We first illustrate the concept of sequential dependencies by means of two examples taken from networks and web services. For each
of these examples, sequential dependencies are extracted and formalized.

2.1   Example 1: Management of Network Services

Computer networks have become over the years the central element of numerous
applications. While they were originally limited to a few basic functionalities
(e-mail, file transfer), a plethora of new services are adding to the value of
these networks: voice over IP, virtual private networks (VPNs), peer-to-peer
communities. The growth of these functionalities over the years considerably
increased the complexity of managing the devices responsible for their proper
functioning, in particular network routers [61].

The deployment of a service over a network basically consists in altering the
configuration of one or many equipments to implement the desired functionalities. We take as an example the Virtual Private Network (VPN) service, which
consists in a private network constructed within a public network such as a
service provider's network [59].

Such a service consists of multiple configuration operations; in the case of
so-called "Layer 3 VPNs", it involves setting the routing tables and the VPN
forwarding tables, setting the MPLS, BGP and IGP connectivity on multiple
equipments having various roles. An average of 10 parameters must be added or
changed in each device involved in the deployment of a VPN.

Figure 1 shows a sequence of commands in the Cisco Internet Operating
System (IOS) for the configuration of a Virtual Routing and Forwarding table
(VRF) on a router. Command `ip vrf` is to be executed first. Its purpose is to
create a VRF instance for the *Customer_1* VPN on a router. This command also
opens the `ip vrf` configuration mode, in which the main VRF parameters can
be configured. These parameters are the *route distinguisher* (`rd`), which allows
the unique identification of the VPN's routes throughout the network, and the
`import/export route target`, which are exchanged between ingress and egress
routers in order to identify the VPNs to which the updates belong to [55, 59].

```
ip vrf Customer_1
rd 100:110
route-target both 100:1000
ip vrf forwarding Customer_1
```

Fig. 1. A sequence of VRF configuration commands for the VPN example

After the creation of the VRF, it is activated on one of the router's interfaces.
However, when done in an uncoordinated way, changing, adding or removing
components or data that implement network services can bring the network in
an inconsistent or undefined state.

In this context, the application of a declarative approach becomes highly desirable. For a network manager, it consists of describing the services or the functionalities available on a network, in conjunction with the modalities for consuming these services or functionalities by the users (customers). To this end, the official documentation for VRF functions in Cisco's operating system [4] describes a clear sequential order of execution among these VRF configuration commands. Firstly, we cannot configure the `rd` and `route-target` parameters of a VRF before creating it with the command `ip vrf`:

Sequential Rule 1. *Command* `rd` *must be called after command* `ip vrf`.

Secondly, we cannot activate a VRF on an interface before creating it (with command `ip vrf`) and configuring its parameters (with commands `rd` and `route target`). Moreover, the VRF we activate must be the one we created, so there is also a relationship between the VRF name in both commands:

Sequential Rule 2. *Command* `ip vrf forwarding` *must be called after commands* `rd` *and* `route-target`. *Moreover, it must have the same argument as* `ip vrf`.

Note that other sequential rules could be similarly expressed for the `route-target` commands and between the `ip vrf forwarding`. The reader is referred to [55] for further details about the setup of a VPN in Cisco's IOS.

2.2 Example 2: Transactional Web Services

As a second example, we take the case of increasingly popular web applications, where some server's functionality is made publicly available as an instance of a *web service* that can be freely accessed by any third-party application running in a web browser. Notable examples of applications using such a scheme include Facebook, Twitter, and Google Maps, among others. In most cases, the communication between the application running in the browser and the web service is done through the exchange of documents formatted in the eXtensible Markup Language (XML), as shown in Figure 2.

```
<message>
  <action>getStockDetails</action>
  <stocks>
    <stock-name>123</stock-name>
    <stock-name>456</stock-name>
    <stock-name>789</stock-name>
  </stocks>
</message>
```

Fig. 2. A sample XML message sent by a web application to a web service

The precise format in which web service operations can be invoked by an applications is detailed in a a document called an *interface specification*; for XML-based interactions, the Web Service Description Language (WSDL) [13]

provides a way of defining the structure and acceptable values for XML requests and responses exchanged with a given service.

While such "data contracts" are relatively straightforward to specify and verify, interface contracts go beyond such static requirements and often include a temporal aspect. For example, the online documentation for the popular Amazon E-Commerce Service [2] elicits constraints on possible *sequences* of operations that must be fulfilled to avoid error messages and transaction failures [39].

As a simple example, we consider the case of an online trading company, taken from [33, 41]. The company is responsible for selling and buying stocks for its customers; it exposes its functionalities through a web service interface. An external buyer (which can be a human interfacing through a web portal, or another web service acting on behalf of some customer) can first connect to the trading company and ask for the list of available products. This is done through the exchange of a getAllStocks message, to which the company replies with a stockList message. The customer can then decide to get more information about each stock, such as its price and available quantity, using a getStockDetails message giving the list of stock names on which information is asked (see Figure 2). The trading company replies with a stockDetails message listing the information for each stock. Finally, the customer can buy or sell stock products. In the case of a buy, this is done by first placing a placeBuyOrder message, listing the name and desired amount of each products to be bought.

This scenario also comprises constraints on the possible sequence of valid operations. For example, a user can call a cash transfer by sending a cashTransfer message without having first specified whether a stock should be bought or sold, or which precise stock is concerned by the transaction. A first choreography constraint would then be:

Sequential Rule 3. *No cash transfer can be initiated before a buy or sell confirmation has been issued.*

A guarantee on the termination of pending transactions can also be imposed. More precisely, the system can require that all transactions eventually complete in two possible ways:

Sequential Rule 4. *Every buy/sell order is eventually completed by a cash transfer or a cancellation referring to the same bill ID.*

Although there exist other constraints in this scenario, these two sequential rules are sufficient for the paper's thesis. For a more thorough study of sequential constraints in transactional web services (and additional examples), the reader is referred to [31, 34].

2.3   Applications of Automated Sequence Generation

The previous scenarios share two important points. First, they involve sets of discrete "events": in the first scenario, events are configuration commands; in the second scenario, events are web service messages sent or received. Second, these events occur in a precise (or at least, not completely arbitrary) *sequence*. Therefore, as we have seen, configuring a VPN not only involves figuring out what commands to issue to a router; these commands will be rejected if they are issued in the "wrong" order. Similarly, interacting with the online trading company will only succeed if the protocol imposed by the web service is respected by both sides.

The prospect of automatically generating sequences of such events, given a set of declarative constraints similar to the previous Sequential Rules, is appealing. As it turns, this concept finds many applications in various fields. We briefly mention a few of them, along with pointers to additional resources for each.

2.3.1 Self-configuration in Autonomic Computing

The automated generation of a structure satisfying a set of properties is part of a more general paradigm called *autonomic computing* [36, 54]; applied in particular to computer networks, this notion is called *autonomic* (or autonomous) *networking* [24].

Self-configuration and self-healing are two central capabilities that a network element must implement in order to exhibit autonomic behaviour [54]. In *self-configuration*, the numerous parameters of a device are populated with appropriate values without immediate external intervention; these values depend both on the element's role and its network context. Narain [51] describes a typical self-configuration architecture, called a *service grammar*. In this context, the requirements language expresses constraints relative to each configuration command; a "synthesis engine" produces commands and appropriate values that fulfil these requirements, which are then applied to each device. The VPN scenario described in Section 2.1 corresponds to this setting: an automatically generated sequence of commands "auto-configures" a VPN.

Most existing configuration management systems, such as Cfengine [12] and Bcfg2 [20], use an imperative rather than declarative approach, and are therefore only partially appropriate for the present problem. An approach suggested by Narain [52] uses a Boolean satisfiability solver to produce a configuration for a set of routers satisfying a number of network constraints expressed in the Alloy language [40]. Such an approach was also explored for the automated generation of configuration parameters in network devices, using different logics, by the authors in [35].

However, none of these solutions involve sequential constraints, and only refer to static snapshots of configurations. As the previous scenarios showed, in many cases finding the appropriate operations and values for these operations is not sufficient.

2.3.2 Automated Composition of Web Services

An open challenge in the web service field is to automatically find a way to make multiple web services interact between each other without a predefined pattern of interaction: this is called *web service choreography* [6]. In such a situation, a script involving all the potential web services is generated dynamically from a set of specifications for each service. It is up to the composition engine to resolve these constraints and call each services' operations in ways that are consistent with each set of requirements.

The main motivation for this setting is the fact that web services forming a choreography can be selected dynamically, at runtime. The classical example is a travel agency, which asks for quotes from various providers before dynamically selecting the best offer and combining reservations details for a car, a hotel room and airline tickets. Clearly, it is impractical to compute in advance all possible combinations of web service providers, and hence their interaction should be guided by the sum of their individual "protocols".

The scenario shown in Section 2.2 corresponds to this setting. The online trading company publicizes a number of data and sequential constraints on its possible invocations. Any potential client must abide by these constraints in order to successfully interact with it.

The automated composition of web services is a very popular research field that has spawned a large number of works over the past decade. Notable contributions include the use of semantic web technologies [53], and in particular *ontologies*, to provide consistent theories of a particular domain intended as a common language for all potential partners. These ontologies can be used, among other things, to describe pre- and post-conditions on operations similar to sequential constraints. Many ontology languages, such as the Ontology Web Language (OWL) [47], use logic as the underlying foundation for expressing their base concepts; automated tasks can then be accomplished on web services annotated with OWL documents using reasoners.

However, the problem can hardly be considered solved, and a consensus has yet to be reached on the way each service's requirement should be expressed, and up to which level of detail. In addition, many ontologies use a variant of first-order logic as their formal language; as we shall see, some properties of first-order logic (namely undecidability) warrant its careful use. Good surveys on existing techniques in web service composition can be found in [17, 57].

2.3.3   Automated Stub Generation

Often a web service under development cannot be run and tested in isolation, and must communicate with its actual partners, even in its early stages of design. Yet for various reasons, it might be desirable that no actual communication takes place. For example, one might want to control the possible responses from the outside environment to debug some functionality; or one might prefer not to provoke real operations (such as buying or selling products) on the actual third-party services during the testing phase.

To this end, a web service "stub" can be constructed. This stub is a kind of mock-up web service, acting as a placeholder for an actual one and simulating its input-output patterns. The degree of "faithfulness" of these mock-ups can vary: sometimes it suffices to return the same response for all inputs of a certain type, while at other times a finer granularity is desirable; the example in Section 2.2 shows that a realistic stub for the online trading company should also simulate transactions involving sequential constraints over multiple operations.

This particular use case of automated sequence generation brings the concept of *interactivity*. Since the stub is composed of inputs sent by a third-party service, and returns responses to these inputs, the sequence of operations is constructed in an incremental fashion. Therefore, at any point in time, an automated generator must not only produce a sequence that satisfies the numerous sequential rules, but must exhibit a trace which is a valid *extension* of a given prefix.

Currently, web service stubs need to be coded by hand, and are hence specific to each development project. Yet, as was pointed in [39], automatically generating such stubs based on a declarative specification of the service would relieve programmers of a task they often overlook. A commercial development tool for web services, called soapUI [3], allows a user to create "mock web services", which consist of hard-coded responses for specific inputs. A similar tool called WebSob [48] can generate random requests and discover incorrect handling of nonsensical data on the service side. However, all these approaches treat request-responses as atomic patterns independent of each other. Of all works, only [31] considers an automated generation of sequences of messages, using a variant of formal grammars.

2.3.4   Test Case Generation

Application testing in a regular operating system can be made by simulating a user generating GUI events. A well-known testing technique, called *monkey testing* [8], involves a testing script generating random sequences of GUI events. However, these events have to be consistent with the state of the application, and hence constraints on the sequence of possible events have to be followed. For example, it does not make sense to send a user interface event for a window that has just been closed: a system crash resulting from that event would not indicate a flaw in the application under test, since the application itself would not allow such a sequence of events to be produced in the first place.

Therefore, an educated monkey tester should generate random sequences of events, taken from the set of all actions that can be produced; as we see, such actions depend on their ordering. An automated sequence generator, provided with a formalization of typical GUI behaviours, provides such a tester. The field of *automated test case generation* concentrates on the production of meaningful test cases for applications under development; in particular, model-based testing attempts to generate these cases from formal models of the system under test. Some approaches use Boolean satisfiability solvers to this end, in the manner of the present paper [43].

3   Automated Sequence Generation through Logic

The previous section has shown how a single concept, namely the automated generation of sequences of operations according to some declarative specification, corresponds to a wide variety of tasks depending on its field of application. In all these fields, the corresponding problem is still considered open. The presentation of related work hence highlights the presence of multiple, seemingly *ad hoc* and domain specific solutions tailored for particular subproblems.

3.1   A Formal Model of Automated Sequence Generation

Yet, the previous scenarios share important common points. They can be modelled as sequences of "events", which contains not only a name, but also a series of parameters and associated values. For example, configuration commands consist of a command name, and of one or more arguments. In the same way, web service messages have an XML structure that can be likened to a command and a list of parameters. From this observation, it is possible to deduce a common formal model, represented in Figure 3.

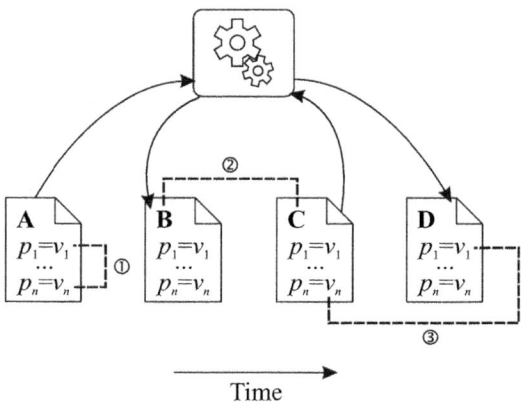

Time

Fig. 3. A formal model to represent declarative constraints in computing. Dashed lines numbered 1, 2, 3 respectively represent static, temporal, and "data-aware" constraints over traces of operations.

The interaction with an object (upper box) is carried by a sequence of operations sent or received (represented by the documents at the bottom). Each of these operations is a tuple $(a, \star)$, where a is a label (represented by letters A-D), and $\star$ is a "content". Formally, let P be a set of parameters, V be a set of values, and A be a set of action names. We define the set of operations as $O \subseteq A \times 2^{P \times 2^V}$ as a set of tuples $(a, \star)$, where $a \in A$ is an action and $\star$ is a relation associating to each parameter $p \in P$ a subset of V.

In the context of network configurations, the operation label can represent the name of a command or of a block of commands to execute; the set of parameter-value pairs represents the parameters of the operation, and in particular the configuration or the portion of configuration the operation acts upon. Previous works have shown that such structures are an appropriate abstraction of configuration information and operations in network devices [30]. For example, in the VPN scenario, we can define these sets as:

$A = \{$`ip vrf`$,$ `rd`$,$ `route-target`$,$ `ip vrf forwarding`$\}$

$P = \{$`vrf-name`$,$ `as`$,$ `send-community`$\}$

$V = \{$`100:110`$,$ `200:210`$,$ `both`$,$ `import`$,$ `export`$,$ `Customer_1`$,$ `Customer_2`$\}$

Using this representation, configuration commands can be represented as operations. For example, the IOS command `neighbor 10.1.2.3 send-community both` can be represented as an operation $o = (\text{neighbor}, \star)$, where $\star$ is the relation such that $\star(\text{address}) = \{10.1.2.3\}$ and $\star(\text{send-community}) = \{\text{both}\}$.

Obviously, some parameters are only appropriate for some operations, and some values are only appropriate for some parameters. For example, the `ip vrf` operation only takes a VRF name, and not a `send-community` parameter. Similarly, it does not make sense for the VRF name to have as a value an IP address such as 10.1.2.3. Therefore, in addition to the previous sets, one can define two ancillary functions:

- A *schema* function $S : A \rightarrow 2^P$ indicating which parameters can appear in an operation for a given action
- A *domain* function $D : P \rightarrow 2^V$ indicating the possible values, among the set V, are allowed for each parameter.

In the VPN example, one can define functions S and D as shown in Table 1.

Table 1. Schemas and domains definitions for the VPN example

$$S = \{ (\text{ip vrf}, \{\text{vrf-name}\}),$$
$$(\text{rd}, \{\text{as}\}),$$
$$(\text{route-target}, \{\text{route-distinguisher}, \text{as}\}),$$
$$(\text{ip vrf forwarding}, \{\text{vrf-name}\})\}$$

$$D = \{ (\text{vrf-name}, \{\text{Customer\_1}, \text{Customer\_2}\}),$$
$$(\text{rd}, \{100\!:\!110, 200\!:\!210\}),$$
$$(\text{send-community}, \{\text{both}, \text{import}, \text{export}\})\}$$

Similarly, in the context of service oriented architecture, the label represents the name of an XML message or of an operation, and the parameter-value pairs represents the content of the XML message. We omit the formalization of the web service scenario, which can be translated in a similar manner as VPN commands.

3.2 Automated Sequence Generation through Satisfiability Solving

A *trace* of operations is a sequence $\bar{o} = o_0, o_1, \ldots$, where $o_i \in O$ for each $i \geq 0$. *Automated sequence generation* can be loosely defined as the algorithmic generation of a trace of operations satisfying a set of constraints expressed declaratively. In the present context, an automated sequence generation architecture can be mapped to some formal concepts. Given:

- some language $\mathcal{L}$ to describe constraints
- a sequence of operations $\bar{o}$ whose values need to be found

- φ, a description of the constraints on $\bar{o}$ expressed in the language $\mathcal{L}$
- P, a procedure that finds values for $\bar{o}$ that comply with φ

then the computation of $P(\varphi, \bar{o})$ finds a sequence of operations that fulfils the constraints. Recall that operations include parameters and values, and that they too must be taken care of if some constraints restrict them. P acts as the synthesis engine, while φ acts as the requirements expressed in the language $\mathcal{L}$.

While there exist multiple ways of solving such a problem, a particularly appealing one is to use logic, or logic-based formalisms, to represent constraints. Rather than providing *ad hoc* algorithms or scripts tailored to a handful of situations, representing each constraint as an *assertion* that must be fulfilled by a potential solution opens the way to more general resolution algorithms. Using logic as the underlying language for expressing dependencies amounts to formulating self-* properties and web service composition broadly as a *constraint satisfaction problem*. Therefore, representing the configuration guidelines in a language $\mathcal{L}$ for which an algorithm P exists allows us to leverage any available CSP tool to solve this problem.

The previous concepts translate easily into logical terminology. A solution $\bar{o}$ consistent with the constraints expressed by φ is called a *model* of φ; we note that fact $\bar{o} \models \varphi$. If the set of constraints expressed by φ admits a solution, φ is said to be *satisfiable*. Procedure P is called a *satisfiability solver*. It need not return a solution for every input. If, for any input φ, the solution $\bar{o}$ (if any) returned by P is such that $\bar{o} \models \varphi$, P is said to be *sound*. Conversely, if any input φ for which there exists a solution $\bar{o}$ is found by P, P is said to be *complete*. A procedure that is both sound and complete computes exactly all the solutions for inputs that have one.

A first advantage of such an approach is its genericity. As long as the language $\mathcal{L}$ is sufficient for expressing all necessary constraints, the procedure P is not tied to any particular instance of a problem and acts as a general problem solver.

A second advantage is that logical formalisms have been thoroughly studied. For example, we shall see in Section 5 a theoretical result about first-order logic which tells us that the problem of finding a finite model for an arbitrary set of constraints is undecidable —that is, unless a tool uses a restricted form of first-order logic, its algorithm cannot be both sound and complete [27, 50]. This kind of global results are seldom available for *ad hoc* methods that do not rest on formal grounds.

Early work on that topic can be traced to Kautz [42], who launched the concept of "planning as satisfiability". However, the present approach differs from traditional planning in some important aspects:

- In planning, transition systems' "states" are atomic objects, and "actions" are represented as binary relations on the set of states. Intuitively, the execution of an action transforms one state into another. Automated sequence generation uses a simplified model where one is simply interested in actions; the internal state of the system is not taken into account, and the focus is on the generation of a sequence of actions satisfying a set of constraints.

- In planning, a "goal state" must eventually be reached. By the previous remark, there cannot be such a "goal" in automated sequence generation; a solution to an automated sequence generation problem is simply a sequence of actions that satisfies the constraints.
- Planning typically aims at computing a complete solution from beginning to end in a single step. Some applications of automated sequence generation are rather interactive (or incremental), as in automated stub generation.

3.3   Desirable Properties for a Logic

The above examples help us determine a few desirable properties that a candidate configuration logic must have in order to be helpful in the scenarios described previously.

First, we can distinguish three types of constraints that can be potentially expressed in the target formal language; Figure 3 provides an example of each.

3.3.1   Support for Static Constraints

A first type of constraint consists of an interdependence relation between multiple data elements inside a single operation or state of the system: we call these dependencies static, which refer to one operation at one moment in time. This relation is described by the dashed line labelled "1" in Figure 3: two parameters in the structure of an operation are constrained by an arbitrary relation. For example, the context could require that parameters p_l and p_n within the same operation have identical (or distinct) values.

To handle this type of constraint, the chosen logic must express relations on parameter-value pairs, such as First-Order Logic [50], which will be described in Section 5.

3.3.2   Support for Temporal Constraints

A second type of constraint consists of an interdependence relation on the sequence of two operations: the constraint is hence called dynamic, since it imposes a relation between two different moments in time. This relation is described by the dashed line labelled "2" in Figure 3; for example, the context could require that the execution of operation "C" always be preceded by the reception of a response labelled "B".

To handle this type of constraint, the chosen logic must express temporal relations. It must include sequencing functions mirroring the temporal operators of logics like Linear Temporal Logic (LTL) [56], which will be described in Section 4.

3.3.3   Support for Data-Aware Constraints

These constraints are the combination of the two previous types of constraints: a relation is imposed on the data content of two operations in two distinct moments in time, as shown by the dashed line labelled "3" in Figure 3. It cannot be considered as a simple static constraint either, since two distinct operations are involved. In the same way, it cannot be considered as a simple temporal constraint, since the content of operations is part of the constraint, and not only the sequencing of these operations.

In the VPN scenario, this is exemplified by Sequential Rule 2. Indeed, this constraint requires that the arguments of all `ip vrf forwarding` and `ip vrf` commands have the same value. This implies an access to the `as` parameter in each such command, and a comparison between values of any pair of them.

In the web service scenario, this is exemplified by Sequential Rule 4. By referring to "every buy and sell order", this new constraint requires that each bill ID appearing in a confirmation eventually appears inside a cancellation or a cash transfer. It implies an access to the data content of the message (the bill ID), and moreover correlates data values at two different moments along the message trace.

This need to access and correlate the data content in multiple messages is not an exception and appears in many other natural properties. Data-aware constraints are therefore more than the sum of their parts; they are best expressed in a "hybrid" variant of both LTL and first-order logic such as LTL-FO$^+$, which will be described in Section 6. The reader is referred to [34] for a detailed description of data-awareness.

In addition to these requirements regarding expressiveness, the target candidate logic should have a number of other properties.

3.3.4   Decidability

To be used as a general constraint solver, the candidate logic must have an algorithm P as described above. This algorithm should be sound, and ideally complete. However, there exist logics for which no such algorithm exists (and will ever exist). Such logics are called *undecidable*. The undecidability of a logic has an important consequence for the task at hand. Suppose that the language used for describing constraints is undecidable. Then any algorithmic procedure P dealing with such a language is *bound* to be imperfect. More precisely, either:

1. It is not sound (and can generate "wrong" solutions) or incomplete (there exist situations where a satisfying assignment exists but P cannot find it); or
2. It cannot be guaranteed to terminate every time —that is, for some inputs, P might fall into an infinite loop.

Note that this consequence stems not from the inability for a programmer to write a correct procedure, but rather from a formal property of the underlying specification language. Automated reasoners for semantic web languages, such as OWL, are often limited by the fact that their underlying language (first-order logic) is undecidable. Remark, however, that decidability can sometimes be restored by imposing restrictions on the way formulæ are written. The loosely guarded fragment of first-order logic, which will be presented in Section 6.2, is an example of conditions on formulæ that guarantees they are decidable.

3.3.5   Finite/Small Model

Another consideration for the logics under study is the size of the solution that their satisfiability algorithm can return. For some logics, there exist sets of constraints whose solutions are infinite. Take for example the statement "for every element x, there exists an element y such that $y > x$". Any model of this formula must have an infinite number of elements (an appropriate model for this expression would be the set of natural numbers, which is infinite).

Clearly, in the present context, one is not interested in solutions involving an infinite number of operations, or an infinite number of parameters for some operation. It is therefore desirable to readily identify sets of constraints leading to such infinite models, or to use logics that prevent the creation of such sets altogether. A logical language that guarantees that if a solution exists, then at least one of them has a finite cardinality, is said to have the *finite model property*. More interestingly, some logics have the stronger *small model property*, where the existence of a solution ensures that at least one of them is not only finite, but bounded by a function of the size of the original set of constraints.

3.3.6   Existence of Automated Solvers

Since the purpose of a logical modelling of sequential constraints is to tap into existing tools for generating models, preference should be given to languages where solvers are available.

In the next sections, we survey various formalisms that could be suitable candidates to express and compute sequences of operations. Each formalism is analyzed with respect to the criteria mentioned above. We shall see that some of them fair differently with respect to each criterion. Yet, any candidate logic for automated sequence generation would be worth studying under these various angles. However, it is not in the paper's scope to express a preference for any of them, as the choice of the "best" language involves a careful compromise between these aspects.

4   Linear Temporal Logic

Linear Temporal Logic [56] is a logic aimed at describing sequential properties along paths in a given Kripke structure. It traces its origins in hardware verification, where it has been widely used to express properties of sequential circuits, among other things [63].

4.1   Description

LTL's syntax starts with *ground terms*, which generally are symbols that can take the value true ($\top$) or false ($\bot$). Such symbols can be combined to form compound statements, with the use of the classical propositional operators $\vee$ ("or"), $\wedge$ ("and"), $\neg$ ("not") and $\rightarrow$ ("implies"). To express sequential relationships, *temporal operators* have been added.

The first such modal operator is $\mathbf{G}$, which means "globally". Formally, the formula $\mathbf{G}\ \varphi$ is true on a given path π when, for all states along this path, the formula φ is true. The second modal operator commonly used is $\mathbf{X}$ ("next"). The formula $\mathbf{X}\ \varphi$ is true on a given path π of the Kripke structure when the next state along π satisfies φ. A formula of the form $\mathbf{F}\ \varphi$ ("eventually") is true on a path when at least one state of the path satisfies φ. Finally, the operator **U** ("until") relates two expressions φ and ψ; the formula $\varphi \, \mathbf{U} \, \psi$ states that 1) ψ eventually holds, and 2) in the meantime, φ must hold in every state. More details can be found in [16].

We must first define the ground terms for our theory:

1. From the set A of actions, we define an equal number of *action terms* α_a for every $a \in A$. On a specific operation $o = (a', \star)$, the term α_a is true exactly when $a = a'$.
2. For every pair of parameter $p \in P$ and value $v \in V$, we create a *tuple term* $t_{p,v}$. On a specific operation $o = (a', \star)$, the term $t_{p,v}$ is true exactly when $v \in \star(p)$.

Equipped with these ground terms, it is first possible to express the schema and domain functions S and D defined in Table 1. For example, since $S(\texttt{ip vrf}) = \{\texttt{vrf-name}\}$, then for any operation whose action is $\texttt{ip vrf}$, only the parameter vrf-name is allowed to have some value. This can be written as:

$$\bigwedge_{p \in P \backslash \{\texttt{ip vrf}\}} \bigwedge_{v \in V} \mathbf{G} \left(\alpha_{\texttt{ip vrf}} \rightarrow \neg t_{p,v} \right)$$

The symbol $\bigwedge_{v \in V}$ is a "meta-symbol", not part of the logic itself. It means that the formula that follows must be repeated once for every value in the set V, by replacing any occurrence of v by that value. For example, if $V = \{k_1, k_2, k_3\}$, then $\bigwedge_{v \in V} t_{p,v}$ would be expanded into $t_{p,k_1} \wedge t_{p,k_2} \wedge t_{p,k_3}$. These meta-symbols can be nested, such that the preceding formula amounts the the repeated conjunction of $\mathbf{G} \left(\alpha_{\texttt{ip vrf}} \rightarrow \neg t_{p,v} \right)$ for all $p \in P \backslash \{\texttt{ip vrf}\}$ and $v \in V$. Hence the previous expression represents $|V|(|P| - 1)$ "copies" of $\mathbf{G} \left(\ldots \right)$, separated by the $\wedge$ connective.

The formula states that every operation in the trace is such that, if the operation's action is $\texttt{ip vrf}$, then every tuple term $t_{p,v}$ must be false whenever p is not $\texttt{ip vrf}$. Such an expression must be repeated for every action and its corresponding schema; this can be formalized generally as follows:

$$\bigwedge_{a \in A} \bigwedge_{p \in P \backslash S(a)} \bigwedge_{v \in V} \mathbf{G} \left(\alpha_a \rightarrow \neg t_{p,v} \right) \tag{1}$$

In the same way, domains for each parameter can be formalized by a similar LTL formula:

$$\bigwedge_{p \in P} \bigwedge_{v \in V \backslash D(p)} \mathbf{G} \, \neg t_{p,v} \tag{2}$$

Sequential constraints in Section 2.1 can now be also expressed as LTL formulæ. For example, Sequential Constraint 1 becomes the following LTL expression:

LTL Sequential Formula 1

$$\neg \alpha_{\texttt{rd}} \, \mathbf{U} \, \alpha_{\texttt{ip vrf}} \tag{3}$$

This formula states that on any trace of operations $\bar{o} = o_0, o_1, \ldots$, action term $\alpha_{\texttt{rd}}$ is not true on any message before action term $\alpha_{\texttt{ip vrf}}$ becomes true. In other words, the action of an operation cannot be rd before some operation has ip vrf as its action; this is indeed equivalent to Sequential Constraint 1.

Similarly, Sequential Formula 2 becomes a slightly more complex expression:

LTL Sequential Formula 2

$$\mathbf{G}\left(\alpha_{\texttt{ip vrf}} \to \bigwedge_{v \in V}\left(t_{\texttt{as},v} \to \mathbf{G}\left(\alpha_{\texttt{ip vrf forwarding}} \to t_{\texttt{as},v}\right)\right)\right) \quad (4)$$

The subformula $\mathbf{G}\left(\alpha_{\texttt{ip vrf forwarding}} \to t_{\texttt{as},v}\right)$ indicates that, at any point in the trace, any operation with action ip vrf forwarding has its as parameter set to some value v. The expression $\bigwedge_{v \in V}(\cdot)$ indicates to take the conjunction of the expression inside the parentheses successively for every value $v \in V$. As a whole, the formula expresses the fact that at any point in the trace, an operation with action ip vrf entails that, for any value v that holds for its parameter as, globally any occurrence of ip vrf forwarding has that same value v for its parameter as.

4.2   Analysis

The conjunction of formulæ (1)–(4) is a (long) LTL formula that provides a complete specification φ of the VRF configuration constraints, expressed in Linear Temporal Logic. It follows that any trace of operations $\bar{o}$ such that $\bar{o} \models_{\text{LTL}} \varphi$ will be a valid sequence of operations, with respect to these requirements.

We now revisit the desirable properties enumerated in Section 3.3, with respect to LTL. First, one can see that, while temporal constraints are easily expressed with LTL's operators, static and data-aware constraints must be simulated through the repeated enumeration of formulæ, changing a Boolean variable each time. Hence, static and data-aware properties become long repetitions of seemingly similar copies of the same formula; more precisely, if $|V|$ is the domain size for values and k is the number of "meta-symbols" ($\bigvee$ or $\bigwedge$) ranging over V, then the length of the resulting LTL formula is in $O(|V|^k)$. Moreover, the expression of that formula must be changed every time the data domain changes.

LTL satisfiability is decidable, and its complexity is PSPACE-complete [25]. As a matter of fact, any LTL formula can be translated into an equivalent *Büchi automaton* [18, 25] whose size is bounded; hence LTL also has the small model property.

While there exist satisfiability algorithms for LTL, which generally involve translating a formula into an equivalent [18, 25], it is also possible to put existing software, called LTL model checkers, to good use. A *model checker* takes as input a finite state machine K, called a Kripke structure, and an LTL specification φ. It then exhaustively checks that any possible execution of K satisfies φ. Popular model checkers for LTL include SPIN [38], NuSMV [14], and Spot [23].

As a by-product of their exhaustive verification, most model checkers produce counter-example traces of a formula, in the case where K does not satisfy φ. It is possible to benefit from this counter-example generation mechanism to find a sequence that does not violate any constraint. It suffices to submit the formula $\neg\varphi$ for verification. If there exists a counter-example for φ, then such a trace must necessarily satisfy φ, giving by the same occasion a valid deployment sequence. It suffices to give the model checker a Kripke structure K where all states are linked to each other, called a universal Kripke structure, and the negation of the constraints as the specification to verify. Such a technique was suggested by Rozier and Vardi [60] and analyzed empirically for a variety of model checkers.

5   First-Order Logic

In Linear Temporal Logic, events are considered as "atomic": they do not contain parameters. To simulate parameters in events, one must therefore resort to create one propositional variable for every parameter and every value, and correctly assign truth values to these variables in every possible operation. Constraints on values become tedious to write, since the same property must be enumerated using every possible value. It would be desirable to have a mechanism where data inside events could be referred to, memorized, and compared at a later time. First-order logic is a formalism that provides such a mechanism.

5.1   Description

In first-order logic (FOL), the traditional propositional connectives $\vee$, $\wedge$ and $\neg$ are used to connect *predicates* $p_1, p_2, \ldots$, which take their values in some arbitrary domain $\mathcal{D}$, and return either $\top$ or $\bot$. The *arity* of predicates is the number of arguments they take; for example, the binary predicate $p(x, y)$ is a function from $\mathcal{D} \times \mathcal{D}$ to $\{\top, \bot\}$.

These predicates and connectives are complemented with a set of two quantifiers. The expression $\exists x : \varphi(x)$ states that there exists some element $d \in \mathcal{D}$ such that φ is true when all occurrences of x are replaced by d. Similarly, $\forall x : \varphi(x)$ asserts the same property, but for all elements $d \in \mathcal{D}$.

In counterpart, the temporal operators and intervals that were available in LTL have disappeared. Temporal relationships between events must therefore be simulated, using additional variables or relations. More specifically, a trace of messages can be represented by a set of three ground predicates:

1. From the set A of actions, we define an *action predicate* $\alpha : \mathbb{N} \times A \to \{\top, \bot\}$. On a specific operation trace $\bar{o} = o_1, o_2, \ldots$, for any $i \in \mathbb{N}$ and $a \in A$, the predicate $\alpha(i, a)$ is true exactly when $o_i = (a, \star)$.
2. From the sets of parameters P and values V, we create a *tuple predicate* $t : \mathbb{N} \times P \times V \to \{\top, \bot\}$. On a specific operation $o_i = (a, \star)$ in an operation trace, for any $p \in P$ and $v \in V$, $t(i, p, v)$ is true exactly when $v \in \star(p)$.

The quantifiers are handy for representing the schema and domain functions. The constraint on the `ip vrf` command, for example, becomes the following first-order formula:

$$\forall p \in P \setminus \{\texttt{ip vrf}\} : \forall v \in V : \forall i \in \mathbb{N} : \alpha(i, \texttt{ip vrf}) \to \neg t(i, p, v)$$

This formula states that, for every pair of parameters (except `ip vrf`) and values p and v, and for every index i, if the action of the i-th operation of the trace is `ip vrf`, then no other parameter-value tuple can be present in the message.

More generally, this expression can be repeated for every action and its corresponding schema, as follows:

$$\forall a \in A : \forall p \in P \setminus S(a) : \forall v \in V : \forall i \in \mathbb{N} : \alpha(i, a) \to \neg t(i, p, v) \tag{5}$$

In the same way, domains for each parameter can be formalized by a similar first-order formula:

$$\forall p \in P : \forall v \in V \setminus D(p) : \forall i \in \mathbb{N} : \neg t(i, p, v) \tag{6}$$

The sequential constraints in Section 2.1 can now be expressed as first-order formulæ. However, special care must be taken to correctly express sequential dependencies as appropriate constraints on the operation indices.

First-Order Sequential Formula 1

$$\exists i \in \mathbb{N} : (\alpha(i, \texttt{ip vrf}) \land \forall j \in \mathbb{N} : (j < i \to \neg \alpha(j, \texttt{rd}))) \tag{7}$$

This formula states that in the operation trace $\bar{o} = o_1, o_2, \ldots$, there exists some operation o_i whose action is `ip vrf`, and such that for every *preceding* operation o_j (where $j < i$), the action for message o_j is not `rd`. This is indeed equivalent to saying that `ip vrf` must precede `rd` in the operation trace.

Similarly, Sequential Constraint 2 can be formalized as follows:

First-Order Sequential Formula 2

$$\forall i \in \mathbb{N} : \forall j \in \mathbb{N} : \forall v \in V : (\alpha(i, \texttt{ip vrf}) \land \alpha(j, \texttt{ip vrf forwarding}))$$
$$\to (t(i, \texttt{as}, v) \leftrightarrow t(j, \texttt{as}, v)) \tag{8}$$

This formula states that for any two operations o_i and o_j in a trace, if o_i has action `ip vrf` and o_j has action `ip vrf forwarding`, then their `as` parameter is equal. The $\leftrightarrow$ operator means "if and only if", and returns true when both its left- and right-hand members have the same truth value. In the present case, this means that for every value $v \in V$, either o_i and o_j both have v as the `as` parameter, or neither does. The global formula is indeed equivalent to stating that the `as` parameter of all `ip vrf` and `ip vrf forwarding` commands is the same for all their occurrences in a trace.

5.2 Analysis

The conjunction of formulæ (5)-(8) is a first-order formula that provides a complete specification of the VRF constraints expressed in first-order logic. Again, any trace of operations $\bar{o}$ such that $\bar{o} \models_{\text{FOL}} \varphi$ will be a valid sequence of operations with respect to these requirements. A model of φ therefore consists in the complete definition of predicates α and t for every $i \in \mathbb{N}$, $a \in A$, $p \in P$ and $v \in V$.

The results of the analysis of first-order logic with the criteria presented in Section 3.3 are almost the opposite as those obtained for LTL in the previous section. First-order logic allows a succinct description of static constraints through the use of universal and existential quantifiers; however, classical FOL does not provide the equivalent of LTL temporal operators. Those must be simulated by explicitly giving constraints on the indices of operations in a trace, stating for which indices a particular formula must hold. This makes the temporal relations much less explicit, as First-Order Sequential Formula 2, for example, shows. Hence, data-aware constraints are also tedious to write, but for a different reason than in LTL.

First-order logic is strictly richer than linear temporal logic: a classical result shows that first-order logic encompasses all of LTL. However, this richness comes at the price of complexity. A result by Trakhtenbrot [62] states that for a first-order language including a relation symbol that is not unary, satisfiability over finite structures is undecidable.

We have seen in Section 3.3 the consequences of undecidability for the constraint language. However, adding constraints to the structure of first-order formulæ might restore decidability. In the particular case where all domains are considered finite and are known in advance, first-order logic statements become decidable [45]. Hence there exist "model finders" for first-order logic that work in that particular case.

Some of the most well-known first-order model finders are Mace [49], Paradox [15], and SEM [64]; they all assume that the cardinality of all domains is bounded. This has an important consequence: it entails that even the length of the trace must be fixed in advance. Hence, in First-Order Sequential Formula 1, the quantifier $\exists i \in \mathbb{N}$ actually becomes $\exists i \in \{0, \ldots, n\}$ for some constant n.

A book by Börger, Grädel, and Gurevich [11] contains an extensive survey on results that assert that certain fragments of first order-logic have a decidable satisfiability problem or not.

6   Linear Temporal Logic with First-Order Quantification

The specification, efficient validation and satisfiability of data-aware constraints is currently an open problem. Although there exists a number of adequate formalisms to specify static and temporal constraints, by the previous observations, they cannot simply be used "side by side" to cover the case of hybrid constraints.

For example, while Linear Temporal Logic allows a succinct formalization of temporal relations, its lack of a quantification mechanism makes the expression of data constraints cumbersome: the same formula must be repeated for every combination of values, replacing them by the appropriate ground terms each time. This results in a potentially exponential blow-up of the original specification, in terms of the domain size.

In contrast, first-order logic's quantification mechanism allows to succinctly express relationships between data parameters between events, but its lack of temporal operators makes the expression of sequential relationships much less natural.

One is therefore interested in striking a balance between the two extremes that these logics provide.

6.1   Description

Linear Temporal Logic with First-Order quantification, called LTL-FO$^+$, aims at this middle ground [32]. As its name implies, LTL-FO$^+$ is a logic where first-order quantifiers and LTL temporal operators can be freely mixed. Its basic building blocks are ground predicates defined as follows:

- From the set A of actions, we define an *action predicate* $\alpha : A \rightarrow \{\top, \bot\}$. On a specific operation o, $\alpha(a)$ is true exactly when $o = (a, \star)$.
- From the sets of parameters P and values V, we create a *tuple predicate* $t : P \times V \rightarrow \{\top, \bot\}$. On a specific operation $o = (a, \star)$, for any $p \in P$ and $v \in V$, $t(p, v)$ is true exactly when $v \in \star(p)$.

One remarks that the predicates are similar to first-order logic, except that they no longer require the "index" specifying at what particular operation in a trace they refer. This, similarly to linear temporal logic, is handled implicitly by the temporal operators.

The Boolean connectives and LTL temporal operators carry their usual meaning. First-order quantifiers, on the other hand, are a limited version of their respective First-Order version. In LTL-FO$^+$, each quantifier is of the form $\exists_p x : \varphi$ or $\forall_p x : \varphi$ and has a subscript, p, which determines which values are admissible for x. More precisely, $\forall_p x : \varphi$ is true for some operation $o = (a, \star)$ if and only if for all values $c \in \star(p)$, φ holds when x is replaced by c. Similarly, $\exists_p x : \varphi$ holds if some value $c \in \star(p)$ is such that φ holds when x is replaced by c. The choice of appropriate values for x therefore depends on the current operation pointed to in the trace; this, in turn, is modulated by the LTL temporal operators.

The development of such a "hybrid" between FOL and LTL has led to many variants of this strategy; notable proponents include LTL-FO [21], EQCTL [44], QCTL [58], Eagle [9], FOTLX [22], and RuleR [10]. However, the particular quantification mechanism presented here is unique to LTL-FO$^+$.

Schema and domain constraints are represented in the same way as for LTL, as is Sequential Constraint 1:

LTL-FO$^+$ Sequential Formula 1

$$\neg\alpha(\texttt{ip vrf}) \, \mathbf{U} \, \alpha(\texttt{rd}) \tag{9}$$

However, using the LTL-FO$^+$ quantifiers, one can represent Sequential Constraint 2 in a much shorter way:

LTL-FO$^+$ Sequential Formula 2

$$\mathbf{G}\left(\alpha(\texttt{ip vrf}) \rightarrow \forall_{\texttt{as}} v : \mathbf{G}\left(\alpha(\texttt{ip vrf forwarding}) \rightarrow t(\texttt{as}, v)\right)\right) \qquad (10)$$

6.2 Analysis

The conjunction of formulæ (1)-(10) is an LTL-FO$^+$ formula that provides a complete specification of the VRF constraints. Again, any trace of operations $\bar{o}$ such that $\bar{o} \models_{\text{LTL-FO}+} \varphi$ will be a valid sequence of operations with respect to these requirements.

As one can see, LTL-FO$^+$ is the formalism where Sequential Formulæ have the shortest expression, compared to both LTL and FOL. Static constraints become similar to first-order logic, while temporal constraints can use the LTL operators. Obviously, data-aware constraints can use both.

However, there is currently no dedicated *satisfiability* solver available for LTL-FO$^+$. Actually, at first glance it is not even clear that LTL-FO$^+$ is decidable, since it includes a form of first-order quantification. However, one can demonstrate that the quantification mechanism of LTL-FO$^+$ is actually similar to the *loosely guarded fragment* of first-order logic [37]. Guarded logic is a generalization of modal logic in which all quantifiers must be relativized by atomic formulæ [7]. If we let $\bar{x}$ and $\bar{y}$ be tuples of variables, quantifiers in the guarded fragment (GF) of first-order logic appear only in the form:

$$\exists \bar{y}\left(\gamma(\bar{x}, \bar{y}) \wedge \psi(\bar{x}, \bar{y})\right)$$

$$\forall \bar{y}\left(\gamma(\bar{x}, \bar{y}) \rightarrow \psi(\bar{x}, \bar{y})\right)$$

The predicate γ, called the *guard*, must contain all free variables of ψ. In this context, it suffices to define the guard as $\gamma(p, x) \equiv \star(p) = y$ and rewrite all LTL-FO$^+$ formulæ as guarded LTL formulæ with general first-order quantifiers:

$$\exists_p x : \varphi \equiv \exists x : (\gamma(x, p) \wedge \varphi)$$
$$\forall_p x : \varphi \equiv \forall x : (\gamma(x, p) \rightarrow \varphi)$$

The advantage of such a rewriting is that the loosely guarded fragment of first-order logic is decidable [26] and has the finite-model property [37]; hence LTL-FO$^+$ is also decidable.

Obviously, LTL-FO$^+$ can be translated back into either LTL or FOL expressions and handled by their respective solvers. Experiments on this concept have led to a first attempt at LTL-FO$^+$ satisfiability solving [28, 29]. This raises the concept that the language used for specification can be different from the underlying representation used by the automated sequence generator.

7 Conclusion

In this paper, we have shown how sequential constraints arise naturally in a variety of computing scenarios, including the management of network devices and the interaction between web services. The automated generation of sequences of operations, following these constraints, amounts to various concepts depending on the field over which it applies. For example, generating sequences of operations corresponds to self-configuration when applied to the management of network device configurations.

The paper argued that such properties be expressed in a declarative manner by means of a formal language, in particular a *logic*. In such a case, the automated generation of sequences becomes a simple instance of *satisfiability solving*, for which solvers already exist in a number of cases.

We first concentrated on two natural candidates for the task at hand, namely Linear Temporal Logic and First-Order Logic. However, we have shown that many of these constraints correlate sequences of operations and parameters inside these operations in such a way that both types of properties cannot be expressed in isolation. This called for the presentation of a family of hybrid formalisms for expressing "data-aware" constraints, a notable proponent being LTL-FO$^+$.

All three logics have been compared with respect to a number of criteria relevant to automated sequence generation. A summary of these results can be found in Table 2.

Table 2. A summary of the results for the logics studied in this paper

	LTL	FOL	LTL-FO$^+$
Static properties	+	++	++
Temporal properties	++	+	++
Data-aware properties	+	+	++
Decidable	Yes	No*	Yes
Small model	Yes	No*	Yes
Existing solvers	+	++	−

The paper focused on a linear temporal–first order characterization of sequential constraints. However, the authors are fully aware that there exist other formalisms that could be suitable for this task. For example, Allen's Interval Temporal Logic [5], Temporal Description Logics [46], interface grammars [31] are all formalisms that should be studied in their own right with respect to the criteria specified in this paper. The presence of such a large roster of possibilities reveals the extent of the uncharted territory on the topic of automated sequence generation with data, as well as its potential for furthering the reach of declarative specifications in computing.

References

1. Proceedings of the 20th Conference on Systems Administration (LISA 2006), Washington, D.C., USA, December 3-8. USENIX (2006)
2. Amazon e-commerce service (2009),
   http://docs.amazonwebservices.com/AWSEcommerceService/2005-03-23/
3. soapUI: the web services testing tool (2009), http://www.soapui.org/
4. Cisco IOS switching services command reference, release 12.3 (2010),
   http://www.cisco.com/en/US/docs/ios/12_3/switch/command/
   reference/swtch_r.html
5. Allen, J.F.: Maintaining knowledge about temporal intervals. Communications of the ACM 26(11), 832–843 (1983)
6. Alonso, G., Casati, F., Kuno, H., Machiraju, V.: Web Services, Concepts, Architectures and Applications. Springer, Heidelberg (2004)
7. Andréka, H., van Benthem, J., Németi, I.: Modal languages and bounded fragments of predicate logic. Journal of Philosophical Logic (27), 217–274 (1998)
8. Arnold II, T.R.: Visual Test 6 Bible. Wiley (1998)
9. Barringer, H., Goldberg, A., Havelund, K., Sen, K.: Rule-Based Runtime Verification. In: Steffen, B., Levi, G. (eds.) VMCAI 2004. LNCS, vol. 2937, pp. 44–57. Springer, Heidelberg (2004)
10. Barringer, H., Rydeheard, D., Havelund, K.: Rule systems for run-time monitoring: From Eagle to RuleR. Journal of Logic and Computation (2008)
11. Börger, E., Grädel, E., Gurevich, Y.: The Classical Decision Problem. Perspectives in Mathematical Logic. Springer, Heidelberg (1997)
12. Burgess, M., Couch, A.: Modeling next generation configuration management tools. In: LISA [1], pp. 131–147
13. Christensen, E., Curbera, F., Meredith, G., Weerawarana, S.: Web services description language (WSDL) 1.1, W3C note (2001)
14. Cimatti, A., Clarke, E.M., Giunchiglia, E., Giunchiglia, F., Pistore, M., Roveri, M., Sebastiani, R., Tacchella, A.: NuSMV 2: An OpenSource Tool for Symbolic Model Checking. In: Brinksma, E., Larsen, K.G. (eds.) CAV 2002. LNCS, vol. 2404, pp. 359–364. Springer, Heidelberg (2002)
15. Claessen, K., Sörensson, N.: New techniques that improve MACE-style model finding. In: Proc. of Workshop on Model Computation, MODEL (2003)
16. Clarke, E.M., Grumberg, O., Peled, D.A.: Model Checking. MIT Press (2000)
17. Daniel, F., Pernici, B.: Insights into web service orchestration and choreography. Int. Journal of E-Business Research 2(1), 58–77 (2006)
18. Daniele, M., Giunchiglia, F., Vardi, M.Y.: Improved Automata Generation for Linear Temporal Logic. In: Halbwachs, N., Peled, D. (eds.) CAV 1999. LNCS, vol. 1633, pp. 249–260. Springer, Heidelberg (1999)
19. Demri, S., Jensen, C.S. (eds.): 15th International Symposium on Temporal Representation and Reasoning, TIME 2008, Université du Québec à Monteéal, Canada, June 16-18. IEEE Computer Society (2008)
20. Desai, N., Bradshaw, R., Lueninghoener, C.: Directing change using Bcfg2. In: LISA [1], pp. 215–220
21. Deutsch, A., Sui, L., Vianu, V.: Specification and verification of data-driven web services. In: Deutsch, A. (ed.) PODS, pp. 71–82. ACM (2004)
22. Dixon, C., Fisher, M., Konev, B., Lisitsa, A.: Practical first-order temporal reasoning. In: Demri, Jensen (eds.) [19], pp. 156–163

23. Duret-Lutz, A., Poitrenaud, D.: Spot: An extensible model checking library using transition-based generalized büchi automata. In: DeGroot, D., Harrison, P.G., Wijshoff, H.A.G., Segall, Z. (eds.) MASCOTS, pp. 76–83. IEEE Computer Society (2004)

24. Gaïti, D., Pujolle, G., Salaun, M., Zimmermann, H.: Autonomous Network Equipments. In: Stavrakakis, I., Smirnov, M. (eds.) WAC 2005. LNCS, vol. 3854, pp. 177–185. Springer, Heidelberg (2006)

25. Gerth, R., Peled, D., Vardi, M.Y., Wolper, P.: Simple on-the-fly automatic verification of linear temporal logic. In: Dembinski, P., Sredniawa, M. (eds.) PSTV. IFIP Conference Proceedings, vol. 38, pp. 3–18. Chapman & Hall (1995)

26. Grädel, E.: On the restraining power of guards. Journal of Symbolic Logic 64(1), 1719–1742 (1999)

27. Grädel, E., Kolaitis, P.G., Libkin, L., Marx, M., Spencer, J., Vardi, M.Y., Venema, Y., Weinstein, S.: Finite Model Theory and Its Applications. Texts in Theoretical Computer Science. An EATCS Series. Springer, Heidelberg (2007)

28. Hallé, S.: Automated Generation of Web Service Stubs Using LTL Satisfiability Solving. In: Bravetti, M., Bultan, T. (eds.) WS-FM 2010. LNCS, vol. 6551, pp. 42–55. Springer, Heidelberg (2011)

29. Hallé, S.: Causality in message-based contract violations: A temporal logic "whodunit". In: EDOC, pp. 171–180. IEEE Computer Society (2011)

30. Hallé, S., Deca, R., Cherkaoui, O., Villemaire, R., Puche, D.: A Formal Validation Model for the Netconf Protocol. In: Sahai, A., Wu, F. (eds.) DSOM 2004. LNCS, vol. 3278, pp. 147–158. Springer, Heidelberg (2004)

31. Hallé, S., Hughes, G., Bultan, T., Alkhalaf, M.: Generating Interface Grammars from WSDL for Automated Verification of Web Services. In: Baresi, L., Chi, C.-H., Suzuki, J. (eds.) ICSOC-ServiceWave 2009. LNCS, vol. 5900, pp. 516–530. Springer, Heidelberg (2009)

32. Hallé, S., Villemaire, R.: Runtime monitoring of message-based workflows with data. In: EDOC, pp. 63–72. IEEE Computer Society (2008)

33. Hallé, S., Villemaire, R.: XML Methods for Validation of Temporal Properties on Message Traces with Data. In: Meersman, R., Tari, Z. (eds.) OTM 2008. LNCS, vol. 5331, pp. 337–353. Springer, Heidelberg (2008)

34. Hallé, S., Villemaire, R., Cherkaoui, O.: Specifying and validating data-aware temporal web service properties. IEEE Trans. Software Eng. 35(5), 669–683 (2009)

35. Hallé, S., Villemaire, R., Cherkaoui, O.: Logical methods for self-configuration of network devices. In: Formal and Practical Aspects of Autonomic Computing and Networking: Specification, Development and Verification, pp. 189–216 (2011)

36. Hinnelund, P.: Autonomic computing. Master's thesis, School of Computer Science and Engineering, Royal Institute of Engineering (March 2004)

37. Hodkinson, I.M.: Loosely guarded fragment of first-order logic has the finite model property. Studia Logica 70, 205–240 (2002)

38. Holzmann, G.J.: The SPIN Model Checker: Primer and Reference Manual. Addison-Wesley Professional (2003)

39. Hughes, G., Bultan, T., Alkhalaf, M.: Client and server verification for web services using interface grammars. In: Bultan, T., Xie, T. (eds.) TAV-WEB, pp. 40–46. ACM (2008)

40. Jackson, D., Schechter, I., Shlyakhter, I.: Alcoa: the Alloy constraint analyzer. In: ICSE, pp. 730–733 (2000)

41. Josephraj, J.: Web services choreography in practice (2005), http://www128.ibm.com/developerworks/webservices/library/ws-choreography/

42. Kautz, H.A., Selman, B.: Planning as satisfiability. In: ECAI, pp. 359–363 (1992)
43. Khurshid, S., Marinov, D.: TestEra: Specification-based testing of Java programs using SAT. Automated Software Engineering Journal 11(4) (2004)
44. Kupferman, O.: Augmenting Branching Temporal Logics with Existential Quantification Over Atomic Propositions. In: Wolper, P. (ed.) CAV 1995. LNCS, vol. 939, pp. 325–338. Springer, Heidelberg (1995)
45. Löwenheim, L.: Uber möglichkeiten im relativkalkül. Math. Annalen 76, 447–470 (1915)
46. Lutz, C., Wolter, F., Zakharyaschev, M.: Temporal description logics: A survey. In: Demri, Jensen (eds.) [19], pp. 3–14
47. Martin, D., Burstein, M., Hobbs, J., Lassila, O., McDermott, D., McIlraith, S., Narayanan, S., Paolucci, M., Parsia, B., Payne, T., Sirin, E., Srinivasan, N., Sycara, K.: OWL-S: Semantic markup for web services (2008), http://www.ai.sri.com/daml/services/owl-s/1.2/overview
48. Martin, E., Basu, S., Xie, T.: Automated testing and response analysis of web services. In: ICWS, pp. 647–654. IEEE Computer Society (2007)
49. McCune, W.: A davis-putnam program and its application to finite first-order model search: Quasigroup existence problems. Technical report, Argonne National Laboratory (1994)
50. Mendelson, E.: Introduction to Mathematical Logic, 4th edn. Springer, Heidelberg (1997)
51. Narain, S.: Towards a foundation for building distributed systems via configuration (2004) (retrieved February 11, 2010)
52. Narain, S.: Network configuration management via model finding. In: LISA, pp. 155–168. USENIX (2005)
53. Narayanan, S., McIlraith, S.A.: Simulation, verification and automated composition of web services. In: WWW, pp. 77–88 (2002)
54. Parashar, M., Hariri, S.: Autonomic Computing: An Overview. In: Banâtre, J.-P., Fradet, P., Giavitto, J.-L., Michel, O. (eds.) UPP 2004. LNCS, vol. 3566, pp. 257–269. Springer, Heidelberg (2005)
55. Pepelnjak, I., Guichard, J.: MPLS VPN Architectures. Cisco Press (2001)
56. Pnueli, A.: The temporal logic of programs. In: FOCS, pp. 46–57. IEEE (1977)
57. Rao, J., Su, X.: A Survey of Automated Web Service Composition Methods. In: Cardoso, J., Sheth, A.P. (eds.) SWSWPC 2004. LNCS, vol. 3387, pp. 43–54. Springer, Heidelberg (2005)
58. Rensink, A.: Model Checking Quantified Computation Tree Logic. In: Baier, C., Hermanns, H. (eds.) CONCUR 2006. LNCS, vol. 4137, pp. 110–125. Springer, Heidelberg (2006)
59. Rosen, E.C., Rekhter, Y.: BGP/MPLS VPNs (RFC 2547) (March 1999)
60. Rozier, K.Y., Vardi, M.Y.: LTL Satisfiability Checking. In: Bošnački, D., Edelkamp, S. (eds.) SPIN 2007. LNCS, vol. 4595, pp. 149–167. Springer, Heidelberg (2007)
61. Tanenbaum, A.S.: Computer Networks, 4th edn. Prentice Hall (2002)
62. Trakhtenbrot, B.A.: Impossibility of an algorithm for the decision problem in finite classes. Doklady Akademii Nauk SSSR (70), 569–572 (1950)
63. von Bochmann, G.: Hardware specification with temporal logic: En example. IEEE Trans. Computers 31(3), 223–231 (1982)
64. Zhang, J., Zhang, H.: System Description: Generating Models by SEM. In: McRobbie, M.A., Slaney, J.K. (eds.) CADE 1996. LNCS, vol. 1104, pp. 308–312. Springer, Heidelberg (1996)

Author Index